Composites Science and Technology

Composites Science and Technology (CST) book series publishes the latest developments in the field of composite science and technology. It aims to publish cutting edge research monographs (both edited and authored volumes) comprehensively covering topics shown below:

- Composites from agricultural biomass/natural fibres include conventional composites-Plywood/MDF/Fiberboard
- Fabrication of Composites/conventional composites from biomass and natural fibers
- Utilization of biomass in polymer composites
- Wood, and Wood based materials
- Chemistry and biology of Composites and Biocomposites
- Modelling of damage of Composites and Biocomposites
- Failure Analysis of Composites and Biocomposites
- Structural Health Monitoring of Composites and Biocomposites
- Durability of Composites and Biocomposites
- Biodegradability of Composites and Biocomposites
- Thermal properties of Composites and Biocomposites
- Flammability of Composites and Biocomposites
- Tribology of Composites and Biocomposites
- Bionanocomposites and Nanocomposites
- Applications of Composites, and Biocomposites

To submit a proposal for a research monograph or have further inquries, please contact springer editor, Ramesh Premnath (ramesh.premnath@springer.com).

More information about this series at http://www.springer.com/series/16333

It-Meng Low · Thamer Alomayri · Hasan Assaedi

Cotton and Flax Fibre-Reinforced Geopolymer Composites

Synthesis, Properties and Applications

Springer

It-Meng Low
Department of Applied Physics
Curtin University
Perth, WA, Australia

Thamer Alomayri
Department of Physics
Umm Al-Qura University
Makkah, Saudi Arabia

Hasan Assaedi
Department of Physics
Umm Al-Qura University
Makkah, Saudi Arabia

ISSN 2662-1819 ISSN 2662-1827 (electronic)
Composites Science and Technology
ISBN 978-981-16-2280-9 ISBN 978-981-16-2281-6 (eBook)
https://doi.org/10.1007/978-981-16-2281-6

This Springer imprint is published by the registered company Springer Nature Singapore Pte Ltd.
The registered company address is: 152 Beach Road, #21-01/04 Gateway East, Singapore 189721, Singapore

Preface

Natural fibres were used in ancient Egypt and China as reinforcement for building materials for pyramids and the great wall. In recent years, due to the increasing environmental awareness, natural fibres have gained popularity by replacing their synthetic counterparts as lightweight reinforcement for inorganic polymers (i.e. geopolymers). The reasons can be attributed to the fibres' relatively low cost and low density, acceptable specific properties, ease of separation, enhanced energy recovery, CO_2 neutrality, biodegradability, and recyclable properties. In addition, natural fibres are durable, reliable, lightweight, and have excellent mechanical properties. For instance, geopolymers have recently become a promising ecological alternative to the traditional cementitious material. They are cost-effective, environmentally friendly, and their production involves relatively small amount of energy. They also have good compressive strength, durability, and thermal properties being highly resistant to flame and heat. However, geopolymers have relatively low tensile and flexural strength, which limits their use in many areas. The mechanical properties of geopolymers can be significantly improved by reinforcement with natural fibres. The resultant geopolymer composites are significantly better than those of traditional materials, and they are fueling the growing demand for natural fibres in various industries such as automotive, building, and construction.

This book provides an overview of the latest advances in the synthesis, characterisation, and mechanical properties of geopolymers reinforced with natural fibres such as pulp fibre, cotton, sisal, flax, and hemp. The influence of adding various natural fibres on the mechanical properties of these composites is discussed. Potential applications, challenges, and future directions of these composites are highlighted and addressed.

Due to the increasing environmental awareness, natural fibre composites based on inorganic polymers (i.e. geopolymers) have aroused great interest among researchers in recent years, and these composites are also becoming more prevalent in use. However, much of the published articles on these materials appeared mostly in journals, and the dissemination of the published work is quite incoherent. There is an impending need to present the work on these materials in a coherent manner. Hitherto, there are no books published on geopolymers reinforced with natural fibres. This book will be the first to provide a comprehensive overview of the latest advances in geopolymer composites with natural fibre reinforcement.

This book consists of seven chapters where background, literature review, and scope of the book are introduced in Chap. 1. The raw materials used and the methodology to fabricate the composite samples by dispersing different types of natural fibres and nano-sized particles into the geopolymeric matrix, together with the test methods to measure the compositions, microstructures, physical, and mechanical properties of samples, are presented in Chap. 2.

Chapter 3 describes the physical properties of geopolymer reinforced with cotton fibres, flax fibres, and/or nanofillers. The effects of these fibres and nanofillers on the microstructure, density and porosity, phase relations, and water absorption are described and discussed. The mechanical and fracture properties of geopolymer composites reinforced with cotton fibres flax fibres and/or nanofillers are described and discussed in Chap. 4. All the mechanical properties of the composites have been improved with the addition of cotton fibres. In general, these composites exhibited improved mechanical properties due to the addition of natural fibres and nanofillers.

In Chap. 5, the effects of natural fibres and nanofillers on the characteristics of moisture absorption and durability of geopolymer composites are presented and discussed. Overall, the durability of these composites has not been affected but improved. I addition, the presence of nanoclay accelerated the process of geopolymerization, reduced the alkalinity of the system, and increased the geopolymer gel. The thermal stability and flammability of these reinforced composites are described and discussed in Chap. 6. With an increase in temperature, the geopolymer composites exhibited a reduction in compressive strength, flexural strength, and fracture toughness. When heated above 600 °C, the composites exhibited a significant reduction in mechanical properties. They also exhibited brittle behaviour due to severe degradation of natural fibres and the creation of additional porosity in the composites.

This book concludes with a summary of major findings on natural fibre reinforced geopolymer composites in Chap. 7 together with some concluding remarks pertaining to challenges and future research directions in this rapidly emerging field.

Perth, Australia — It-Meng Low
Makkah, Saudi Arabia — Thamer Alomayri
Makkah, Saudi Arabia — Hasan Assaedi

Contents

About the Authors

Prof. It-Meng Low gained his B.Eng. and Ph.D. degrees in Materials Engineering from Monash University prior to taking up lecturer positions first at Auckland University and then Curtin University. In 1986–1988, he conducted post-doctoral research with Prof. Y-W. Mai on fracture and toughening micromechanics of epoxy systems at Sydney University. He was awarded a visiting professorship by the Japanese Ministry of Education to work with Prof. Niihara at Osaka University in 1995/1996. He is a fellow of the Australian Ceramic Society and serves on the editorial board of several journals. He is also the recipient of the prestigious 1996 Joint Australian Ceramic Society/Ceramic Society of Japan Ceramic Award for excellence in ceramics research. He has authored/edited more than 12 books and is a author of over 250 archival research papers. His h-index is 45 with more than 6250 citations. His research interests include polymer composites, nanotechnology, toughening, and failure micromechanics. He is the current WA Branch President and Federal Councillor of the Australian Ceramic Society. He is also the recipient of the prestigious 1996 Joint Australasian Ceramic Society/Ceramic Society of Japan Ceramic Award for ceramics research.

Dr. Thamer Alomayri gained his Ph.D. degrees in Materials Science from the Department of Imaging and Applied Physics at Curtin University, Australia, in 2015. He serves as a reviewer for several journals. He is author of over 29 research papers. His h-index is 13 with more than 614 citations. His research interests include material science research, development of geopolymer composites (GC), application of nanoparticles, natural fibre in geopolymer, and polymer nanocomposites. He currently works as an associate professor at Faculty of Applied Science, Department of Physics at Umm Al-Qura University, Makkah, Saudi Arabia.

Dr. Hasan Assaedi is working as an assistant professor in the Department of Physics at Umm Al-Qura University. He gained a bachelor's degree in physics from King Abdulaziz University in 2002. He received his M.Sc. in Theoretical Physics from the University of Adelaide, Australia, in 2012, and then he obtained his Ph.D. in Physics from Curtin University, Western Australia, in 2017. His research focuses on developing sustainable and environmentally friendly materials such as geopolymers reinforced with ductile fibres, and he also has research in developing optical properties of semiconductors. He has 32 peer-reviewed publications including one chapter in a book. His h-index is 8 with 484 citations according to Google Search.

Chapter 1
Background

Abstract This chapter provides a background to the scope of this book and a comprehensive review of the latest advances in geopolymer composites reinforced with natural fibres and nanofillers. The influence of adding various natural fibers and nanofillers on the mechanical properties of these composites is reviewed. The synthesis, structure and properties of these reinforced composites and nanocomposites are also described. Potential applications, challenges, and future directions of these composites are highlighted.

1.1 Introduction

1.1.1 Overview of Composite Materials

Composite materials are multiphase materials that consist of at least two distinct components. One of their major attractions is their capacity to optimise the properties of their materials because combining the constituents makes possible interactions that give rise to properties that the constituents on their own cannot achieve. In most cases the primary two constituents are usually distinguishable: one, referred to as the matrix, is generally the dominant material; the second component is typically referred to as the reinforcement.

One of the practical benefits of composites is that there are almost unlimited combinations of matrices and reinforcements. For manufacturers, this means that it is possible to tailor the properties of composites to meet the requirements of specific applications. Typically, manufacturers aim to develop composites that feature improvements in mechanical properties: enhanced strength, stiffness, or toughness, for instance. Other relevant considerations are often the weight of the material, its durability, its thermal stability, and especially its cost. In some settings, the last is increasingly being re-defined to encompass not only private and economic but also public and environmental costs.

I.-M. Low et al., *Cotton and Flax Fibre-Reinforced Geopolymer Composites*, Composites Science and Technology,
https://doi.org/10.1007/978-981-16-2281-6_1

Composite materials are commonly classified based on their matrices. Ceramic matrix composites (CMSs), polymer matrix composites (PMCs), and metal matrix composites (MMCs) are arguably the three main structural composites. While all these groups can offer composites with high strength and stiffness, excellent chemical and thermal stability, and relatively low density, one weakness can be material brittleness. The issue is particularly relevant for ceramic matrix composites. Ceramic materials alone have an inherently brittle nature, and this considerably limits their suitability for use in engineering and structural applications.

One way to overcome the brittleness is by adding reinforcement. Fibre is a common reinforcement material because of its capacity to stabilise micro-cracking, the formation of tiny cracks on either side of a main crack front. The incorporation of short fibre and continuous fibre are the two leading techniques comprising ceramic matrix reinforcement. Of these, continuous fibres are more expensive and more troublesome to mix into matrices; however, their direction of alignment provides superior mechanical properties of the composites. Continuous fibre composites are often used in aviation applications where the benefits of enhanced properties can be fully exploited. In contrast, short fibres are cheaper to develop and easier to produce. Short fibre ceramic composites are well established with regard to applications where lower strength and stiffness are sufficient.

The constituent materials' properties go a long way to influence the properties of the resultant composite materials. Most importantly, the type, amount, distribution and orientation of reinforcement and void content of constituent materials determine the mechanical and other properties of the composite materials. The nature of the interaction between matrix and the reinforcing fibres and load transfer mechanisms at the interface are also important determinants of final composite material properties. To better ensure superior mechanical properties, it is important that a strong bond is formed between the fibres and the matrix. The presence of such bonding enables effective transfer of stress from the dominant matrix to the reinforcing fibres. There are also different techniques to improve the adhesive quality and strength properties of fibre-reinforced composites, such as the addition of minerals and chemical treatment of fibres. The method of processing also influences the properties of composites, because the processing techniques affect the amount and dispersion of fibre within the resultant composite. Thus, researchers and developers are very interested in the role of processing conditions and techniques on performance optimisation.

1.1.2 Fibre-Reinforced Geopolymer Composites

While ceramic matrix composites have been increasingly studied over the past four decades, their commercial relevance is still small in relation to that of polymer matrix composites and metal matrix composites. One reason for this is the high production cost of ceramic matrix composites, which tend to require materials that are costly and manufacturing temperatures that are extremely high. Thus far, the high cost of producing ceramic matrix composites has made them undesirable for particular commercial applications. To overcome this issue, there has been renewed interest in identifying novel inorganic, non-metallic materials and techniques that can be used to produce high-strength and cost-effective ceramic matrix composites.

Geopolymers, more formally known as inorganic aluminosilicate polymers, offer an alternative. These materials differ from more common ceramic matrix materials in terms of their favourable density and thermal stability, featuring temperature resistance up to 1000 °C. Geopolymers are typically produced using aluminosilicate sources reacting with an alkali activation solution maintained at high pH and near-ambient temperature. Geopolymer binders are generally viscous at the outset. When cured they harden in a manner like organic thermosetting resins. The result is a three-dimensional solid material which is generally amorphous and has ceramic-like properties. The reasonable mechanical properties of geopolymers, and low temperature processing, make a unique combination offering considerable potential in the field of cost-effective inorganic composite development (Davidovits 2002); thus, the use of geopolymers as a replacement for conventional ceramic matrix composites is an important area of interest. Another area of interest is the use of geopolymer technology as a precursor for true ceramic matrix composite fabrication due to the typical transformation of geopolymers into crystalline ceramic phases at higher temperatures.

Davidovits in 1987 started investigating the application of geopolymers as matrix materials (Davidovits 2002). The rationale for research at the time was that fibres could reinforce geopolymers to form composites for use in moulding patterns and tools for the industry of plastic processing. It has been found that the incorporation of fibres indeed does significantly improve a geopolymer's mechanical properties. Basalt, carbon and glass fibres are the most commonly used, but there are disadvantages in such synthetic fibres (Beckermann 2007). The first of these is their abrasive character. This means that they increase wear on machinery and are generally dangerous to work with. Another disadvantage is that synthetic fibres, including glass fibres, are troublesome to dispose of legally and with consideration for the environment once their life ends. These fibres, because of their tendency to result in residues that cause furnace damage, cannot be incinerated. They cannot be recycled as reprocessing operations tend to cause fibre breakages and other associated problems. To date, disposal to landfills is the dominant option for discarding this waste, but this is a costly option when considered on a large scale, given government duties associated with landfill services (Beckermann 2007).

One response to the disadvantages of synthetic fibres is to re-consider the use of fibres more compatible with the environment. Banana, coir, cotton, flax, hemp, jute, sisal, and wood are examples of natural fibres that meet this description (Sreenivasan et al. 2011). These fibres, unlike many plastics, are biodegradable, meaning that they will not remain on the surface of the earth for thousands of years. Furthermore, natural fibre production uses less energy than carbon or glass fibre production (Venkateshwaran et al. 2011). Natural fibres have lower densities (1.25–1.5 g/cm^3) than the densities of E-glass (2.54 g/cm^3) or carbon fibre (1.8–2.1 g/cm^3); they are also lighter than synthetic fibres (Sgriccia et al. 2008; Anuar and Zuraida 2011). Another important property of natural fibres is their excellent modulus–weight ratio, which makes them particularly useful for stiffness-critical designs. Natural fibres have been investigated and used in the automobile interior product design industry for some time because of their superior acoustic damping properties. For purposes of noise attenuation, natural fibres are generally superior to carbon or glass (Mallick 2007). In terms of mechanical properties, natural fibres tend to possess excellent specific modulus, toughness, flexibility, and specific strength properties (Bax and Müssig 2008; Monteiro et al. 2009). Increasingly it has been noticed that natural fibres tend to be significantly more commercially viable than most synthetic fibres (Dhakal et al. 2007).

Another important benefit of all plant-derived natural fibres is that their initial growth in plant form is dependent on the consumption, not emission of CO_2. This means that these fibres are substantially CO_2 neutral materials. Practically, this means that incineration at their life's end is not considered to release additional carbon dioxide into the atmosphere. In contrast, glass fibres and other synthetic fibres are not attributed to CO_2 neutral status and require that fossil fuels be burned in order for energy to be generated for their production. Recent socio-political attentions focusing on carbon emissions have meant that most commercial practices rely on the burning of fossil fuels have come under some level of scrutiny. As carbon dioxide emission has been increasingly associated with the greenhouse effect and climate change, social and economic sanctions concerning carbon dioxide emissions are arguably likely to increase before they decrease in the near future.

Natural fibre-reinforced composites feature the benefit of not splintering during fracture. This distinguishes them from most of their glass fibre-reinforced counterparts and means that they are more suitable for uses where safety is a concern, such as in crash absorption applications. The durability of natural fibres is another advantage. These fibres can generally be recycled and reused with a relatively minimal reduction in strength and stiffness properties. In contrast, synthetic fibres tend to fracture as a result of recycling and further processing, resulting in reduced fibre length.

Cotton fibre is amongst the most well-known of the natural fibres. It offers excellent absorbency, a natural feel and other properties of comfort. It can be distinguished from most other natural fibres by its cellulose content, which is very high (90–95%) (Andre 2006). While this fibre has been used in the textile industry for centuries, it has more recently attracted interest as a reinforcement material for composites. Reclaimed cotton fibre is widely used as a cheap to fill composites used as interior parts in the automotive industry (Müssig 2008). In 2003, it is estimated, 45,000 tonnes of material were used in the German motor vehicle industry for interior products alone (Karus et al. 2005). Cotton fibre, in a similar fashion to other natural fibres, is a lightweight, bio-degradable and eco-friendly material. The advantages of cotton fibres are that they are not as expensive to source and are not as brittle as carbon fibres (Chaudhary and Gohil 2013). While the weight of cotton fibres is preferable for automobile applications because it reduces the weight of the overall motor vehicle, other advantages are their ability to reduce cabin noise through sound wave attenuation (Bhat et al. 2004). Cotton fibres are also used in commercial applications: they have been used to reinforce gypsum composites producing a high-quality building material. According to Li et al. (2003), the cotton fibre/gypsum composite features low density, favourable thermal properties, acoustic insulation, and a very high strength-to-weight ratio.

Many studied have been reported recently on the physical, thermal and mechanical performance of cotton fibre-reinforced polymer composites. Hashmi et al. (2007) found that the addition of 27.5 vol.% cotton fibres to an unsaturated polyester resin matrix composite increased the impact strength of the composite, per unit width, from 61 to 971 Nm/s^2; it also increased the flexural strength from 101.8 to 142 MPa. The modulus of elasticity at bending was found to have increased from 2.4 to 4.2 GPa. Fervel et al. (2003) carried out studies on polyester composites reinforced with cotton fibres. The team investigated friction and wear of the composite against stainless steel. The result was that as the volume fraction of cotton fibre increased in the composite, the coefficient of friction also increased. A similar result was found with respect to wear: as the volume fraction of cotton fibre increased the rate of wear was found to decrease. This was true to 15 vol.% and after: that is, at greater volume fractions of cotton fibre, the rate of wear was found to be substantially constant. Fervel et al. (2003) also studied the effects of fibre orientation on the cotton fibres/polyester composites. It was found that after initial variations, and after the friction coefficient had stabilised, the orientation of fibres influenced this coefficient. Hashmi et al. (2006) found that when cotton fibres were added to 33 wt% to an unsaturated polyester resin matrix, the structural integrity of the composite improved with respect to sliding wear. The team found that the improvement in sliding wear conditions gave a composite with structural integrity twice as good as un-reinforced polyester resins in this regard.

Hendra and Peer Mohamed (2010) investigated the influence cotton fibre addition (10–30 wt%) on the impact strength, tensile strength, flexural strength and flexural modulus of unsaturated polyester resin matrix composites. They found that each of these properties increased markedly as cotton fibre content increased to 30 wt%, with tensile strength reported at 65.66 MPa, flexural strength at 180 MPa, flexural modulus at 7.1 GPa and impact strength at 37.5 kJ/m^2. Raftoyiannis (2012) investigated the use of cotton fibre as a reinforcement for composite panels to be used in construction. It was found that the mechanical properties of these cotton fibre reinforced composites were suitable for the composites to be used in doors and other secondary structural materials. It was also concluded that the structural performance of these cotton fibre composites was satisfactory for building products with low requirements, including wall panels.

Rukmini et al. (2013) studied the effect of cotton fibre addition on the mechanical properties of polypropylene (PP) composites. The team found that the addition of 30 wt% cotton fibre caused the tensile strength and tensile modulus of the PP composite to increase substantially. While pure PP was found to have a tensile strength of 21.87 MPa and a tensile modulus of 618 MPa, the composite was found to give results of 28.07 and 1867 MPa, respectively. The improvements in flexural strength and flexural modulus were also notable, from 21.7 to 45.3 MPa, and from 813 to 1925 MPa. The team concluded that cotton fibre reinforcement was an effective technique for enhancing the mechanical properties of PP composites.

To date, while there have been a few investigations into the impact of cotton fibre in polymer matrices, there has been no research into the influence of cotton fibre addition on the mechanical properties of geopolymer composites. Given the context of increasing interest in more environmentally sustainable materials and practices, this present work seeks to address this gap in the prevailing research by exploring the impact of cotton fibre addition on the mechanical properties of geopolymer composites.

1.2 Literature Review

1.2.1 Geopolymers

(i) *Overview*

The use of cementitious materials is central to the construction industry. Ordinary Portland Cement (OPC) remains the most commonly applied cementitious material in the sector. In manufacturing OPC; however, large amounts of resources are used such as energy and natural materials. Another concern is the use of OPC contributes to adverse environmental consequences, studies show that a huge amount of greenhouse gases is released into the atmosphere during production. It takes approximately 1.5 tons of natural materials to produce an estimated 1-ton OPC. The process of creating the single ton of OPC releases an estimated 1 ton of CO_2 into the atmosphere. Researchers suggest that 6–7% of the total CO_2 emitted worldwide originates to the cement industry (Davidovits 1994a; Chen et al. 2014; Shaikh 2013). Besides the potential environmental concerns related to the production of OPC, the substance also has drawbacks in relation to resistance to chemicals and limited mechanical strength in a number of applications (Chen et al. 2014).

Geopolymers, also known as inorganic polymers, have attracted attention as a potential alternative for OPC in the construction industry. Geopolymer provides an alternative form of cementitious material that is produced from a rich source of aluminosilicates. Chemically, aluminosilicates create a reaction when an alkaline solution is applied with temperatures at either slight elevation or ambient ranges. The types of raw materials that are used in the production of geopolymer are readily obtainable through natural sources (Shaikh 2013; Liew et al. 2016).

An abundant supply of by-products that can be used in the production of geopolymer is available. The resources that may be used in the production of geopolymer may include industry by-products, volcanic ash or meta-kaolin. The by-products that may be used include furnace slag, fly-ash or mine tailings. The industrial by-products created by a number of industries has become a problem due to the difficulty in disposal solutions. Geopolymers may provide a solution for disposal while providing more enhanced performance when compared to the traditional form of OPC. However, there are additional benefits than the reduction of greenhouse gases and waste. The advantages for the use of geopolymer include using a material which is fire and acid resistant, hazardous materials and toxins can be immobilized, rapid curing and adherence to a number of aggregates. (Part et al. 2015; Chen et al. 2014).

(ii) *Geopolymer Molecular Model and Chemical Reaction*

Davidovits first coined the term "geopolymer". The term was first used to refer to aluminosilicate polymers that had an amorphous microstructure. The researcher recommended the chemical designator for geopolymer made of aluminosilicates the term polysialate (Davidovits). Sialate is represented as the abbreviation in silicon-oxo-aluminate when the alkali is sodium (Na^+) or potassium (K^+). Polysiliate is referred to as chain and ring polymers. Polysiliates are polymer molecules include chains of Si and Al ions in four fold coordination linked with oxygen, and contains a variety between amorphous to semi-crystalline (Davidovits 1994b). The empirical formula of geopolymer molecules is written as:

$$Mn\left(-(SiO_2)_z - AlO_2\right)_n, wH_2O \tag{1.2.1}$$

M represents the cation, *n* represents the degree of polycondesation and z is equal to 1, 2, 3 and higher. The geopolymer molecules are observed as molecules with amorphous to crystalline aluminosilicates structures. They can be classified based upon the ratio (Si:Al) to polysialate, polysilate-siloxo and polysialate-disiloxo (Davidovits 2008).

A simplified model of the reaction process in geopolymer was offered by Duxson (2007). The researcher indicated that the significant part of the processes starts when the aluminosilicate specimen dissolves when exposed to highly alkaline environments (Duxson et al. 2007). Once the aluminosilicate source material is exposed to an alkaline environment, the dissolution process began immediately, which resulted in breaking the covalent bonds that connect aluminium, silicon and oxygen molecules. The molecules then polymerize into a structured gel for a short period. The results reveal a three-dimensional chain polymeric structure made from Si–O–Al–O bonds. The structured gel is formed when the oligomers create a network during the aqueous phase by releasing water through condensing and dissolution process. The newly formed gel will reorganize from gel 1 to gel 2 as more water is released. Finally polymerization, through a stage of condensation occurs (Duxson et al. 2007a).

Another model was proposed by Fernández-Jiménez et al. (2005) who considered a framework for dissolution of particles of fly-ash in an alkaline environment. The research findings revealed that the pH level set by the activator system had a high influence on the rate of dissolution and activation of fly-ash. The sphere of fly-ash starting the dissolution process as a visible part of the shell has begun to dissolve. The dissolution process will continue as the alkaline liquid begins to penetrate the interior of the sphere and begin to create the dissolution process from the inside out. The process causes the formation of aluminosilicate gel, a product of the process, in and on the sphere of fly-ash. The gel which forms in larger fly-ash spheres can block the penetration of the alkaline liquid. The blocked alkaline material is then unable to penetrate further reaction on smaller particles and are observable when the dissolution process is complete. The size of the fly-ash particles and variations in pH levels cause a uniform dissolution results in the gel.

(iii) *Geopolymer Synthesis*

Geopolymers are constituted by two parts. The first part includes materials that provide a reactive aluminosilicate solid (raw materials), similar to what is found in fly-ash or metakaolin. The second part is the solution that activates the alkaline reaction of either alkali metal hydroxide or silicate (Liew et al. 2016).

(iv) *Sources of Aluminosilicates*

An aluminosilicate is a raw material which is used to make geopolymers. The selection of raw materials must be critical in order to achieve the desired level of performance from geopolymers (Liew et al. 2016). The reason that aluminosilicates are chosen for the process is in the rich amount of alumina (Al_2O_3) and silica (SiO_2) available in the material. Alumina and silica are an abundant resource found inside the earth's crust.

The aluminosilicate source materials, however, are preferred to be in a reactive phase (Cioffi et al. 2003). The best raw materials to use in the manufacturing of geopolymers are those rated high in Silica and Alumina while in amorphous form (Williams et al. 2011). The polymerization reaction greatly depends on the quality of the raw materials used in the process. The quality of the materials is important as the structural properties of geopolymers depend on its purity.

By-products and Mineral raw materials have been studied and reported as a foundation material for producing geopolymer. For example, metakaolin has been studied extensively as a source material of geopolymers (Palomo et al. 1999a; Perera et al. 2005; Duxson et al. 2007b; He et al. 2012; Williams et al. 2011). Investigations into low-calcium fly-ash as a material showed a number of potentials (Bakharev 2005, 2006; Jun and Oh 2015). A number of studies are available on the results of using natural Al and Si minerals (Xu and Van Deventer 2000). Other research has suggested using a combination of calcined as well as non-calcined mineral (Xu and Van Deventer 2002). Researchers have also proposed using a combination of metakaolin and fly-ash (Van Jaarsveld et al. 2002). Materials such as combining rice husk ash (Detphan and Chindaprasirt 2009) and furnace slag (Cheng and Chiu 2003; Islam et al. 2014) have also been offered as alternatives to be used in the creation of geopolymers. Each of the materials which have been offered and studies have been found to contain high amounts of reactive and amorphous alumina and silica.

- Metakaolin

In the early years of studying raw materials to create geopolymer synthetics, metakaolin was a commonly investigated material (Davidovits 1991). However, in recent years, fly-ash has become the most popular raw material due to the applications offered in construction. Metakaolin is obtained through a process known as calcination or the dehydroxylation when kaolin clay is heated to temperatures ranging from 600 to 800 °C (Ferone et al. 2013; Kong et al. 2007). The process removes the chemically bonded molecules of the water and the octahedral coordinated aluminum.

The octahedral coordinated aluminum is discovered within the kaolin in a four to five fold configuration (Palomo et al. 1999a).

- Volcano Ash

Volcano ash has also been studied to use as a raw material to make geopolymer. Volcano ash is made up of minerals, rock and small volcanic fragments of glass created through the volcanic process. Volcano ash contains high amounts of silica and alumina. However, the crystalline phases were found high in the volcanic ash, so the ash has been mixed with other aluminosilicates sources such as metakaolin and fly-ash in order to produce better geopolymers (Tchakoute Kouamo et al. 2013).

- Rice Husk Ash

As the title suggests, rice husk ash is made from burning rice husks. When rice husks are burned, the lignin and cellulose are consumed by the heat and leave behind a substance high in silica ash. Rice husk ash (RHA) is found to contain high amounts of amorphous silica. The amorphous silica is a highly porous convex structure. The material is found to be a successful silica additive in the creation of geopolymers (He et al. 2013; Rattanasak et al. 2010; Bohlooli et al. 2012; Detphan and Chindaprasirt 2009).

- Fly-Ash

Fly-ash refers to the raw material derived from the industrial by-product made from generating power through the combustion of coal. When coal is burned within a combustion chamber, ash begins to form on the bottom of the chamber (bottom ash) and fine particles rise up the flue with the hot gases. The fine particles which rise with the hot gas are collected by filtration devices such as electrostatic precipitators prior to reaching the chimney.

The Standard Specification for Coal Fly-ash and Raw or Calcined Natural Pozzolan for use in Concrete or the ASTM C 618-15, classifies fly-ash in three category types. Each type is based upon the major chemical substance (Table 1.1). The SiO_2, Al_2O_3 and Fe_2O_3 must meet the standard requirements such as 70% in class N and class F. Class C shows a minimum amount requires SiO_2, Al_2O_3 and Fe_2O_3 to be between 50 and 70%.

Table 1.1 The chemical requirements of fly ash according to ASTM C618-15

Requirements	Class		
	N	F	C
Silicon dioxide (SiO_2) plus aluminum oxide (Al_2O_3) plus iron oxide (Fe_2O_3), min, %	70.0	70.0	50.0
Sulfur trioxide (SO_3), max, %	4.0	5.0	5.0
Moisture content, max, %	3.0	3.0	3.0
Loss on ignition, max, %	10.0	6.0	6.0

Incombustible material amounts, and types of the coal determine the fly-ash chemical composition after the combustion process. Bituminous and anthracite coals produce Class F fly-ash while Class C is formed through burning lignitic and sub-bituminous coal. Specifications of calcium contents are not stated from each class. However, Class C fly-ash calcium contents is expressed as calcium oxide (CaO), which have found to be higher than Class F. Class F fly-ash is found to contain pozzolanic properties. While Class C contains both cementitious and pozzolanic properties.

Over the last decade, new research has focused on creating fly-ash geopolymers with lower amounts of calcium (Palomo et al. 1999b; Duxson et al. 2007a). Generally, class F fly-ash with low-calcium is considered the preferred material when compared to Class C fly-ash with higher levels of calcium. The high presence of calcium may be found to interfere in the polymerization process by altering the microstructure (Gourley 2003; Shaikh 2013). However, a research found that fly-ash contains high levels of CaO to form greater compressive strength. The increase of compressive strength has been found to occur because of the development of compounds such as calcium-aluminate-hydrate (Van Jaarsveld and Van Deventer 1999).

Fly-ash from coal is considered a variable material in terms of the physical and chemical proprieties of the particles. The reactivity and the chemistry of fly-ash particles depends on variables such as the source coal, the pre and post-combustion conditions (Kutchko and Kim 2006). However, research revealed that classified fly-ash from given source locations exhibits some constancy over a period of time. Fly-ash phase composition changes depending on the power plants the material is sourced from. The source fly-ash will react in similar fashion as long as there are no changes to coal source or burning conditions. Fly-ash's bulk chemical composition may be quantitatively determined using X-ray fluorescence (XRF). Tables 1.2 and 1.3 show chemical composition and estimated particle's sizes of five Australian types of fly-ash, respectively (Gunasekara et al. 2014).

Table 1.2 Chemical composition of fly-ash collected from five different locations in Australia (Gunasekara et al. 2014)

Fly-ash type	By weight (%)										
	SiO_2	Al_2O_3	Fe_2O_3	CaO	K_2O	TiO_2	P_2O_5	MgO	Na_2O	SO_3	LOI
Gladstone (GFA)	50.82	29.89	10.26	3.24	0.58	2.05	1.61	0.80	0.00	0.28	0.43
Port Augusta (PAFA)	49.97	31.45	3.22	5.03	1.87	2.54	1.77	1.54	1.85	0.33	0.51
Collie (CFA)	52.67	29.60	11.27	0.94	0.65	1.83	1.13	0.72	0.00	0.48	0.63
Mount Piper (MPFA)	65.18	25.30	1.90	0.63	3.65	1.53	1.21	0.00	0.00	0.23	1.30
Tarong (TFA)	73.12	21.50	1.36	0.29	0.63	1.84	1.06	0.00	0.00	0.00	1.16

Table 1.3 Particle size distribution of fly-ash collected from five different locations in Australia (Gunasekara et al. 2014)

FA type	Passing (%)										Surface area (m^2/kg)
	10 µm	20 µm	30 µm	40 µm	45 µm	50 µm	60 µm	70 µm	80 µm	90 µm	
GFA	43.1	61.9	73.2	79.8	82.7	85.3	89.6	91.2	92.6	93.8	2003
PAFA	46.7	62.1	71.4	77.4	80.9	82.9	87.9	90.1	92.1	93.8	2161
CFA	40.9	54.6	62.7	67.7	70.0	72.3	76.7	79.0	81.3	83.6	1934
MPFA	36.0	57.1	69.9	77.4	80.7	83.8	89.0	91.2	93.0	94.6	1555
TFA	43.0	63.0	73.6	79.3	81.8	84.2	88.3	90.2	91.9	93.4	1766

Fly-ash is primarily amorphous. Beside the amorphous content, class F fly-ash has crystalline phases made from quartz, magnetite, mullite and hematite (Rattanasak and Chindaprasirt 2009; Lee and van Deventer 2002; Rickard et al. 2011). The crystalline content within fly-ash show stability through the alkali activation process and do not dissolve during geopolymerization. However, the crystalline content may still affect the properties of the geopolymer.

X-ray diffraction (XRD) is used to determine the phase composition of fly-ash. Diffractometer and search-phase software can be used in the lab to perform qualitative analysis of data. An internal standard such as fluorite (CaF_2) usually is mixed with fly-ash powder before analysis in order to determine the amorphous and crystalline content in fly-ash quantitatively. XRD quantitative analysis usually is conducted using software that use Rietveld refinement modeling (Williams et al. 2011; Rickard et al. 2011). Chen-Tan et al.analyzed Colie fly-ash quantitatively and concluded that the only content that is reactive in fly-ash is the amorphous aluminosilicates during the geopolymerisation reaction (Chen-Tan et al. 2009). Therefore, to define the reactivity of any fly-ash type, phase composition analysis must be conducted. Table 1.4 shows the phase composition of fly-ash collected from three different resources as presented by Rickard et al. (2011).

(V) *Activator Solution*

Alkaline solutions are critical. Such solutions are necessary to activate the source materials in each geopolymerisation process. Sodium hydroxide (NaOH) is used most often as an alkali activator when combined with sodium silicate (Na_2SiO_3). Another alkali activator is potassium hydroxide (KOH) with potassium silicate (K_2SiO_3). A singular activator may also be functional in geopolymeric reaction.

Table 1.4 Phase composition of three different class-F fly-ash (Rickard et al. 2011)

Phase	Formula	Collie FA (wt%)	Eraring FA (wt%)	Tarong FA (wt%)
Amorphous content		54.00 (45)	62.74 (31)	50.82 (28)
Mullite (ICSD 66452) Mullite (ICSD 66449)	$Al_{4.56}\ Si_{1.44}\ O_{9.72}$ $Al_{4.59}\ Si_{1.41}\ O_{9.7}$	15.80 (18)	20.88 (14)	25.1 (11)
Quartz low (ICSD 83849)	SiO_2	11.14 (18)	8.08 (16)	10.31 (14)
Quartz low primary (ICSD 83849)	SiO_2	15.05 (21)	6.81 (14)	13.77 (13)
Magnetite (ICSD 43001)	$Fe_3\ O_4$	2.51 (83)	1.491 (52)	
Hematite (ICSD 88417)	$Fe_2\ O_3$	1.50 (64)		

For example, successful research has shown synthesizing geopolymer using a single sodium hydroxide activator with rice husk ash and red mud (He et al. 2013).

The type and concentration of each alkali activator is a main factor in geopolymerization process (Komljenović et al. 2010). The researchers observed five different types of alkali activators. The first alkali activator used in the research was calcium hydroxide ($Ca(OH)_2$). The second activator used was sodium hydroxide (NaOH). Sodium hydroxide was observed while combined with sodium carbonate (Na_2CO_3). Potassium hydroxide (KOH) and Sodium silicate (Na_2SiO_3) was also examined as useful alkali activators. A number of different concentrations were examined to create fly-ash geopolymer. The curing conditions were maintained at a constant control to accurately study the alkali activators effect regarding the mechanical properties. Sodium silicate showed the strongest compressive strength over each of the tested alkali activators (Komljenović et al. 2010). The next strongest was calcium hydroxide, sodium hydroxide. Sodium hydroxide combined with sodium carbonate and potassium hydroxide. The activation potential of potassium hydroxide was low when compared to sodium hydroxide; this difference was attributed to the size difference of sodium and potassium ions. The study also concluded that the optimum value of sodium silicate modulus was estimated to be 1.5 (Komljenović et al. 2010). It was therefore suggested that higher levels than the standard modulus may result in a loss in compressive strength of the matrices. The study also revealed that compressive strength values depended greatly on the concentration of alkali activators. Higher compressive strength results have been shown with higher concentrations of all types of activators.

Geopolymerization is a process that greatly depends on the use of alkaline solutions. Strong alkaline solutions are needed to increase surface hydrolysis of aluminosilicate particles (Hu et al. 2009). The alkali concentration is an important factor when dissolving Si and Al during the geopolymerisation process; the amount of ions leached in the process is mostly dependent on the concentration of the alkali activator. Research suggests that enhancement in compressive strength can be achieved by increasing the concentration of alkali activators (Somna et al. 2011; Ahmari and Zhang 2012; Gorhan and Kurklu 2014; Yusuf et al. 2014; Hanjitsuwan et al. 2014).

1.2.2 Microstructural Properties of Gopolymers

Geopolymers phase composition is typically assessed with X-ray diffraction (XRD). XRD phase patterns of geopolymers commonly show both crystalline and amorphous phases. Amorphous phase is found at a broad peak of $2\theta = 20$–$30°$. Sharp crystalline phases can be found in both sources of aluminosilicates and geopolymers, which means that the crystalline contents do not react during the geopolymerization process. According to Rattanasak and Chindaprasirt (2009), mullite and quartz are found in geopolymer samples of crystalline fly-ash bases. They concluded that the silicate phase in geopolymerisation processes was highly chaotic (Rattanasak and Chindaprasirt 2009).

Fourier Transform Infrared Spectroscopy (FTIR) is currently being used by researchers to discover the reaction product analysis of geopolymer materials (Rees et al. 2008; Perná et al. 2014; Phair and Van Deventer 2002; Gao et al. 2014). FTIR patterns give molecular data of the chemical bonds found in geopolymers as each compound presents a different vibration. Different wavelengths and different intensities provide information of molecular bonds formations in geopolymer pastes. Thus, FTIR techniques may give evidence of effective geopolymerization practices.

Features of the FTIR spectra can be distinguished by the bands at ~1000 cm^{-1}. The band is a representation of a Si–O–Si tetrahedron which is a typical band in geopolymers (Chindaprasirt et al. 2009; Arioz et al. 2013; Gao et al. 2014). Therefore, some researchers used the highest and the area under the peak Si–O–Si vibration as indicator to the degree of geopolymerisation (Chindaprasirt et al. 2009; ul Haq et al. 2014). Based on shape and location of the Si–O–Si band, Chindaprasirt et al. (2009) argued that the concentration of alkaline solution is one of the main factors in geopolymerization. The molecular band Si–O–Al may also be lowered in frequency during polycondensation alternating the Si–O and Al–O bonds (Lee and van Deventer 2002). Bands found approximately at 3450 cm^{-1} were observed to be for O–H stretching, while 1650–1600 cm^{-1} was ascribed to O–H bending (Zaharaki et al. 2010; Rattanasak and Chindaprasirt 2009).

Scanning electron microscopes (SEM) may also be used to examine geopolymer's microstructure and precursors. Geopolymers microstructures are usually examined using SEM to provide an image of defects, cracks, morphology, porosity and reactions of aggregates. It is common to find unreacted particles within the geopolymer gel. Geopolymers with developed dissolution reaction includes high proportions of gel comparing to unreacted fly-ash particles (Kriven et al. 2003). Pores and micro-cracks are commonly found in geopolymer matrices as well.

1.2.3 Thermal Properties of Geopolymers

The thermodynamic processes of geopolymers are typically measured using, differential thermal analysis (DTA), differential thermogravimetry (DTG), and thermogravimetric analysis (TGA). Each technique assists in gathering information on phase stability and thermodynamic properties of materials (ul Haq et al. 2014; Kong and Sanjayan 2010; Li et al. 2012; Duxson et al. 2007b). To study thermal stability and eliminate oxidation reaction, nitrogen or argon may be used as an inert atmosphere. Strong thermal resistance has been discovered in geopolymers, such resistance may be useful as industrial or domestic insulation.

Due to their thermal properties, the automotive and construction industry has discovered a number of uses for geopolymer (Liefke 1999). Liefke (1999) suggests the use of foamed geopolymer for insulation in a number of areas. One characteristic in geopolymer that is studied extensively is thermal conductivity. Thermal conductivity is affected by different parameters such as the material density, porosity, chemical composition and fillers. Materials that are low in density and high in porosity are applicable to be a good insulator, as the air that filled the microstructural voids has low conductivity.

1.2.4 Mechanical Properties of Geopolymers

The mechanical properties of geopolymers are variable. Such properties are dependent on each geopolymer's relative amounts of aluminium, alkali, silicon, and water. Solid to liquid ratio or amount of amorphous Al_2O_3:SiO_2 and Si:Al are significant binder variables. Ratios including silicon and aluminium are important as they are directly affected the molecular network of geopolymer. Another important factor in geopolymer formation quality is curing conditions; this includes some sub factors such as curing temperatures and duration of curing. This section presents the mechanical properties with consideration of geopolymer binder variables.

(i) *Effect of Activator Settings on Mechanical Properties of Geopolymers*

Alkali concentration is important in geopolymer production. The concentration of alkali activator solutions impacts on the mechanical properties of geopolymer samples. Features of concern when using alkaline solutions is when sodium silicate is combined with sodium hydroxide. Ratios between Sodium silicate and sodium hydroxide (Na_2SiO_3:NaOH) are essential to the geopolymerization process. By changing sodium silicate to sodium hydroxide ratio, the strength and the economy aspects of geopolymers can be optimized (Yusuf et al. 2015). Additionally, the effect of the activator modulus (Ms) should also be studies. Ms is defined as the mass ratio from silicon dioxide (SiO_2) to sodium oxide (Na_2O). This section will outline effects observed of alkali concentration, Na_2SiO_3:NaOH and Ms in relation to the mechanical strength of geopolymers.

Alkali materials promote and speed up the geopolymerzation process. The concentration of alkali depends greatly on the ion numbers and pH levels. By increasing the concentration of sodium hydroxide, the metakaolin geopolymer strength is also increases (Alonso and Palomo 2001; Mishra et al. 2008). However, a number of researchers believe high levels of alkalinity are unfavorable to the strength of geopolymers. The increased strength observed when increasing concentration in sodium hydroxide showed a sharp decrease after reaching optimum levels. Sodium hydroxide may disrupt leaching Si and Al ions from the aluminosilicate at high concentrations. This may lead to premature precipitation in geopolymer gel and a reduction in mechanical strength of the material (He et al. 2013).

Rice husk ash is also used for the creation of geopolymer. The greatest strength volume has been identified by researchers was at a concentration of 12 M sodium hydroxide (Nazari et al. 2011). Khale and Chaudhary presented a research in the relation between samples pH and strength exhibited. Samples of pH 14 revealed over 50 times higher results than pH 12. Their review concluded that pH range between 13 and 14 is considered to be ideal in the creation of geopolymer that is high in mechanical strength (Khale and Chaudhary 2007). Studies have shown that higher in caustic alkalinity causes more successful dissolution of raw aluminosilicates and activate higher amounts of alumina and silica, which consequently increased geopolymer gel and the strength of matrices (He et al. 2012).

The compressive strength of fly-ash-geopolymer in term of sodium hydroxide concentration while cured at the ambient temperature was reported (Somna et al. 2011). The concentration of sodium hydroxide was changed between 4.5 and 16.5 M. A rapid improvement in compressive strength was observed when the concentrations increased up to 14 M (Somna et al. 2011). The observed increased strengths were discovered to be a result of the high leaching of alumina and silica species. However, the compressive strength of samples prepared with sodium hydroxide concentrations at 16.5 M appeared to decrease. The excess of hydroxide ions caused precipitation in the aluminosilicate gel, which resulted in lower strength geopolymers (Somna et al. 2011).

A study presented by Gorhan and Kurklu (2014), examined the compressive strength of fly-ash geopolymer with different sodium hydroxide concentrations over a 7-day experiment. All samples were thermally cured for 2, 5 and 24 h at 65 and 85 °C temperatures. Concentrations of 3, 6 and 9 M of sodium hydroxide was used in the preparation of samples. The ideal sodium hydroxide concentration for the greatest compressive strength was found 6 M that achieved 21.3 MPa and 22 MPa for samples cured at 65 °C and 85 °C, respectively, for 24 h (Gorhan and Kurklu 2014). This optimal concentration gave an alkaline atmosphere to dissolute the source material without hindering the polycondensation process. The concentration at 3 M is considered to be too low and unable to stimulate strong reactions while the high concentration at 9 M caused premature coagulation in silica leading to a weaker geopolymers (Gorhan and Kurklu 2014).

Ahmari and Zhang studied the mechanical performance of geopolymer created from copper mines and sodium hydroxide. The concentrations of sodium hydroxide varied between 10 and 15 M. The variations were used to understand the effect of sodium hydroxide concentration on the unconfined compressive test of geopolymers. The unconfined compressive strength determined that samples with 15 M concentration were higher than 10 M samples. The variation in strength was found to be caused by the higher sodium hydroxide-to-aluminosilicates ratio which caused, in turn, higher Na:Si and Na:Al (Ahmari and Zhang 2012).

When creating geopolymers, it is important to pay attention to the ideal ratios of sodium silicate to sodium hydroxide solutions Na_2SiO_3:NaOH when designing geopolymers. Sodium hydroxide solution becomes a dissolvent while the sodium silicate becomes a binder during the reaction. According to Ridtirud et al. (2011), the optimum ratio of Na_2SiO_3:NaOH in fly-ash based geopolymers was 1.5. Ratios studied were 0.33, 0.67, 1.0, 1.5 and 3.0. The highest compressive strength value recorded was 45.0 MPa in the case of Na_2SiO_3:NaOH = 1.5. Increase in strength was attributed to the sodium content. The Na^+ ion is critical to geopolymer formation due to charge balancing ions during the reaction process. Although, excessive silicate throughout the reaction process caused a reduction in strength because of delay in water evaporation during the reaction which made an interference in geopolymer molecules creation (Ridtirud et al. 2011).

Sathonsaowaphak et al. (2009) reported the effect of liquid alkaline:ash ratio, Na_2SiO_3:NaOH ratio and sodium hydroxide concentration on geopolymer's compressive strength. It was found that a liquid alkaline:ash ratio of 0.4209–0.709, a Na_2SiO_3:NaOH ratio of 0.67–1.5 and a sodium hydroxide concentration of 10 M respectively, led to superior compressive strength and ideal workability in geopolymer samples. The use of 10 M sodium hydroxide was important, since sodium hydroxide solution improves the dissolution of silica and alumina's species, and sodium ions act as a charge balancing (Sathonsaowaphak et al. 2009).

Salih et al. (2014) found that in the case of geopolymer produced using palm oil fuel ash, the optimum Na_2SiO_3:NaOH ratio was 2.5 and the optimum solid:liquid ratio was 1.32 for maximum compressive strength. The researchers found that solid:liquid ratios of less than 1.32 led to a higher presence of voids which adversely affected compressive strength. Similarly, Na_2SiO_3:NaOH ratios higher than 2.5 led to excessive amounts of sodium silicate which hindering the geopolymerization reaction.

Sukmak et al. (2013) investigated the influence of liquid:fly-ash and Na_2SiO_3:NaOH ratios on the development of compressive strength of geopolymers created using clay and fly-ash as aluminosilicates source materials. Liquid:fly-ash ratios applied were 0.4, 0.5, 0.6 and 0.7, and the Na_2SiO_3:NaOH applied were ranging between 0.4 and 2.3. The outcome was that liquid:fly-ash ratios <0.3 and >0.8 failed to be suitable for clay-fly-ash geopolymers due to null strength. Liquid:fly-ash and Na_2SiO_3:NaOH ratio optimum values were 0.6 and 0.7 respectively. This result was lower than the optimum ratio in the case of fly-ash-based geopolymers. This is may be attributed to the fact that clay has a higher tendency to absorb the cations and is likely to absorb added sodium hydroxide. The greatest compressive strength achieved was roughly 15 MPa at 90 days of curing (Sukmak et al. 2013).

Activator modulus (Ms) is a variable that can determine the soluble silicate amount used in geopolymers. It controls the dissolution rate as well as the gelation and polycondensation throughout the chemical reaction of geopolymer. Thus, it impacts on the ultimate strength development of geopolymer material. A suitable Ms must be designed and chosen depending on the chemical composition of the raw aluminosilicate materials.

Law et al. (2014) reported that 1.0 is an optimum Ms for fly-ash geopolymer, and any subsequent increase fails to result in compressive strength increase. It was suggested that greater than 1.0 Ms, all particles of fly-ash had dissolved or that there was an absence of further dissolution of each fly-ash particle (Law et al. 2014). Yusuf et al. (2014) reported that the influence of activator modulus Ms on the strength of samples produced from ground steel slag and palm oil fuel ash was small. In the study, compressive strengths obtained were 69.1 MPa and 65.0 MPa by a modulus of 0.915 and 1.635, respectively (Yusuf et al. 2015).

Komljenović et al. (2010) investigated the influence of Ms on fly-ash geopolymers mechanical properties. With increasing sodium silicate activator modulus, the ratio Si:Al of the reaction products increased, and Na:Si and Na:Al decreased. The researchers based on the results of compressive strength, found that greater compressive strength was associated with greater modulus values, and greater Si:Al ratio of geopolymers (Komljenović et al. 2010).

Guo et al. (2010) investigated the activator modulus and alkali activator content (Na_2O %) influence on the compressive strength of fly-ash based geopolymers. Combinations of Na_2SiO_3 and NaOH were applied as activator. The alkali activator's modulus was increased from 1.0 to 2.0 and alkali activator content was ranging from 5.0 to 15%. The alkali activator content was founded based on the mass quantity of Na_2O to fly-ash. Silica-alkali modulus and alkali activator content were found to be critical in geopolymers strength development. The content of alkali activator (Na_2O %) and the optimum modulus were recorded as 10% and 1.5% respectively. This produced compressive strength values that were recorded to be 22.6 at 3 days, 34.5 at 7 days and 59.3 MPa at 28 days when the samples were cured at ambient temperature (Guo et al. 2010).

The previous studies investigated factors that are relevant to the activator and procedure of geopolymer preparation. However, most of the aluminosilicate contents involved in the geopolymeric reaction is derived from the source material. It is necessary to determine the amorphous contents of the source materials since they have an essential role in forming the resultant geopolymers.

(ii) *The Effect of Molar Ratios on Mechanical Properties of Geopolymers*

The sodium content in a geopolymer system is usually given by sodium hydroxide and sodium silicate solutions. The aluminosilicate sources and sodium silicate collectively are responsible for the silicon content while the aluminosilicates alone are responsible for the aluminium content in geopolymers. Sodium hydroxide and sodium silicate liquids as well as free water that were added during the mixing process are responsible for the water content. Furthermore, the various different mixing parameters and ratios i.e. the solid:liquid ratio, Na_2SiO_3:NaOH ratio and the NaOH concentration are responsible for the differences in oxide molar as well as the atomic ratio in a geopolymer system. Nevertheless, the quantity in which each component is used in a geopolymerization reaction extensively depends on the reactive phases or the reactivity of aluminosilicates. It is quite common in the published studies to figure out the molar ratios of geopolymers such as Si:Al by considering the chemical composition of the precursor materials. However, this is may be inaccurate since the crystalline aluminosilicate phases are unreactive through the geopolymer reaction. The actual molar ratios of geopolymer gel can be determined experimentally using quantitative XRD analysis (Rickard et al. 2011, 2015), quantitative EDS Analysis (Rowles and O'Connor 2003) or both (Williams et al. 2011).

The effective Si:Al ratios affects the dissolution as well as the hydrolysis and the polycondensation process of geopolymers. Furthermore, raising the ratios of SiO_2:Al_2O_3 can improve the mechanical properties of the geopolymer (Davidovits 2008). The content of silica is also seen to have a great influence on the mechanical properties of geopolymers. Also, the alumina content of the geopolymer controls its setting. This can be one of the reasons for the increase in the dissolution of aluminosilicates in high concentration of Si content in the geopolymerization reaction (Dimas et al. 2009; Palomo et al. 1999b).

Rowles and O'Connor (2003) examined the impact of the ratios Na:Al and Si:Al on the compressive strength of metakaolin geopolymers activated by sodium silicate and sodium hydroxide. They discovered that both varying ratios seemed to have a considerable impact on the measured compressive strength of geopolymer. They provided a contour map which showed the compressive strength values for geopolymers in term of the ratios Na:Al and Si:Al as presented by the authors. The optimum strength (53.1 MPa) was obtained by studying the geopolymers with a ratio of 1.29 in Na:Al and 2.50 with Si:Al. Furthermore, it was also recorded that the ratios were based on the source material measurement rather than the geopolymer gel. Since the entire metakaolin did not react, the actual ratio of Na:Al and Si:Al of the geopolymer gel produced would not have been the same as the calculated ones. Amongst the amorphous geopolymer gel, unreacted aluminosilicates remain as a secondary phase (Rowles and O'Connor 2003).

The degree of geopolymerization of the dissolved species is controlled by the SiO_2:Na_2O molar ratio. A rise in the content of Na_2O was seen as a responsible factor in improving the strength gain in the geopolymers, an increase in the setting rate as well as a significant reduction of cracking in the final product (Xu and Van Deventer 2002).

Gao et al. (2014) investigated the influence of SiO_2:Na_2O ratio on metakaolin based geopolymers. On the basis of the results obtained by the experiment, it was seen that the setting time of the metakaolin-based geopolymer increased with the SiO_2:Na_2O ratio due to the viscous property of the sodium silicate. When the ratio of SiO_2:Na_2O in geopolymers was recorded as 1.50, the sample showed less porosity, and thus better compressive strength (Gao et al. 2014). In yet another study carried out by Soleimani et al. (2012), metakaolin geopolymer was manufactured with different Na_2O:SiO_2 activator ratios ranging between 0.3 and 1.1 and cured at room temperature for 1 and 4 weeks. According to this study, when the sample's ratio of Na_2O:SiO_2 was raised up to 0.6, the strength of the samples increased and recorded 32 MPa after 4 weeks of curing (Soleimani et al. 2012).

The compressive strength of geopolymers can also be improved by optimizing the SiO_2:Al_2O_3 ratio in geopolymers. One way of optimization is to combine two different aluminosilicate source materials in order to adjust the ratio (He et al. 2013; Yan and Sagoe-Crentsil 2012). Nazari et al. (2011) proposed a new and innovative way to alter the chemical composition of the resulting geopolymer by mixing recycling husk-bark ash, a high silica source, with fly-ash. Various concentrations of NaOH i.e. 4, 8 and 12 M as well as sodium silicate were used as chemical activators for the purpose of stabilizing the mix. Here, the ratio of Na_2SiO_3:NaOH and the chemical activator to solid source material was fixed to be at 2.5 and 0.4. Husk-bark-ash was loaded to the mix at varying amounts (20, 30 and 40 wt%). Following that, the samples were left for 24 h for pre-curing, then they were exposed to oven curing at 80 °C for at least 36 h. In the end, the authors concluded that a rise was observed in the compressive strength of the blended husk bark ash and fly-ash geopolymer at almost every single fly-ash replacement level. Furthermore, at fly-ash replacement of 30% and any concentration of sodium hydroxide solutions, showed the greatest values of compressive strength in the different geopolymer matrices being studied. The strength values of all mixes were found to range somewhere between 20 and 30 MPa (Nazari et al. 2011).

The alkalinity of alkali reactant solution can be expressed in the form of Na_2O:H_2O. It has been reported that even though Na_2O:H_2O does not affect or alter the nature of the final product, the ratio holds considerable importance (Rahier et al. 1997). It has been observed that an increase in Na_2O:H_2O ratio can be held responsible for enhanced dissolution ability as well as the mechanical strength development within various clay-based geopolymers (Xu and Van Deventer 2000) According to Latella et al. (2008), the low content of water (H_2O:Na molar ratio less than 5.5) were responsible for cracks formation in the sample after 10 days of curing. Furthermore, higher amounts of porosities were seen developing in geopolymers with the molar ratio H_2O:Na = 6. However, when the molar ratio equals to 5.5, the geopolymer matrices gained the highest bulk density among other samples. Table 1.5 shows the bulk density and open porosity of metakaolin geopolymer samples in term of H_2O:Na molar ratio as tested on the 7th day after creating. (Latella et al. 2008).

(iii) *Effect of Water Content on Mechanical Properties of Geopolymers*

When it comes to the alkali activation reaction, the process of geopolymerization typically includes a reaction between the dissolved species of alumina and silica. This chemical reaction takes place in a highly alkaline environment. In a geopolymer system, water simply provides a medium of transportation between the dissolved alumina and silica ions (Yunsheng et al. 2010). Furthermore, water can also give workability to the freshly prepared pastes since it does not contribute to the reaction directly (Chindaprasirt et al. 2007; Jansen and Christiansen 2015).

Still, the addition of water during the formation of the geopolymer is always seen with concern, since it is seen to be responsible for the dilution of alkalinity of the system as well as the transportation of ions away from the reaction zone (Barbosa et al. 2000; Zuhua et al. 2009). Since the entire reaction is a water releasing process, it may thwart the process of geopolymerization (Davidovits 2008). According to Rahier et al. (2007), either too high or too low concentration of water content in the reaction can be held responsible for decreasing the geopolymerization process. The amount of OH-ions is affected by the water content in the geopolymeric reaction, which in turn affects the efficiency of the chemical reaction (Rahier et al. 2007).

Table 1.5 Open porosity and bulk density values in term of the molar ratio H_2O:Na for metakaolin based geopolymers

H_2O/Na molar ratio	Open porosity (%)	Bulk density (g/cm^3)
2	6	1.85
4	12	1.64
4.5	16	1.55
5.5	20	1.57
6	30	1.30

Apart from it all, water is seen to have a direct effect especially on the open porosity as well as the density of the geopolymer matrices. More often than not, higher level of water contents results in an increase in the total porosity of the geopolymer (Zhao et al. 2009; Lizcano et al. 2012; Latella et al. 2008). This causes a decrease in the mechanical strength of geopolymer matrices.

Geopolymer's water content, as Ahmari and Zhang (2012) and others found, significantly affects performance and mechanical strength. They looked at how initial water contents effected UCS (or unconfined compressive strength) in copper mine-tailing geopolymer. More specifically yet, they created their test geopolymer samples at 6 different water content levels between 8% (the lowest) and 18% (the highest), each separated by 2% water content. The initial water contents were calculated as the ratio (by mass) between water in sodium hydroxide activating solutions and solid contents in the mixture. Those geopolymer pastes then were placed inside steel molds, whereby they were compressed until a saturation state was achieved. The samples were then tested after one week. Results showed that as the initial water contents increased, UCS increased in all geopolymer samples (Zhao et al. 2009). The unconfined compressive strength of 33.7 MPa was resulted at an original water content of 18% (and 0.2 foaming pressure). Such improvement in the mechanical performance of geopolymers was ascribed to the role water plays during geopolymerization as liquid medium (Ahmari and Zhang 2012). Similar results were found in other experimental work. Compressed autoclave bricks of alkali-activated material and low-level silicon tailings (both slag and fly-ash) reached a peak strength of 16.0 MPa at water percentages ranging between 6.5 and 8.0 wt%. Higher amounts of water content, however, resulted in decreased compressive strength (Zhao et al. 2009).

Water dependency is also determined by other mixing parameters related to the alkali activator and raw materials in use. Water contents in geopolymers, for instance, need to also consider solid:liquid ratio, alkali concentration, as well as alkaline reactant ratios. Komljenović et al. (2010) noted potentially significant effect in water:fly-ash ratios to geopolymers strength, though the activators used also factored. In general, the geopolymer's compressive strength increased with decreased water:fly-ash ratios. In potassium hydroxide activated geopolymers, however, the geopolymers displayed lower strength at low water:fly-ash ratios. Revealing low activation potential in the case of potassium hydroxide when compared to other activators (Komljenović et al. 2010).

Also noteworthy is how source material fineness also affects water demand. For instance, metakaolin requires higher liquid demands compared to fly-ash, due to differing particle shapes. Metakaolin's shape is that of structured layers, whereas fly-ash particles are spherical. As the layered structure limits particle mobility in mixing, it is consequently less workable, and metakaolin geopolymers require lower solid:liquid ratios when compared to fly-ash for homogeneous mixing. According to Kong et al. (2007), the ideal solid:liquid ratios for fly-ash and metakaolin geopolymers were 3.0 and 0.80, respectively. Beyond aforementioned ratios, pastes were found to lose workability in both cases (Kong et al. 2007).

(iv) *Curing Conditions Effect on Mechanical Properties of Geopolymers*

Better durability and mechanical performance require suitable curing temperatures (Komnitsas and Zaharaki 2007). Heat increases reaction rates by accelerating dissolution from aluminosilicates into alumina and silica species, and thus facilitates polycondensation and geopolymer paste hardening (Alonso and Palomo 2001; Sathonsaowaphak et al. 2009; Kong et al. 2008). Thus, heat is required to initiate the geopolymerization reaction by surpassing the reaction's thermal activation point. Evidently, if temperatures are very high, or the exposure time is too long, the process can also cause a weakening in the material strength (Pangdaeng et al. 2015).

As a result, many attempts to study varied curing temperatures ranging from room temperature to 120° for geopolymerization reactions have been reported (Giasuddin et al. 2013; Aydin and Baradan 2012; Ahmari et al. 2012; Ridtirud et al. 2011; De Vargas et al. 2011; Bakharev 2006; Mishra et al. 2008).

Palomo et al. (1999) reported that geopolymer matrices when cured for 24 h at 85 °C gained much greater compressive strength than when cured at 65 °C, though if curing time extended to more than 24 h, increased strength was much less (Palomo et al. 1999b). Other research found that curing metakaolin geopolymers at ambient temperatures proved unfeasible, though increased temperatures (40 °C through 100 °C) resulted in strength gains after curing 1–3 days; curing for longer durations or at higher temperatures, however, caused samples to later fail (Heah et al. 2011). As Rovnaník (2010) noted, that metakaolin-based geopolymers are cured at higher temperatures (40–80 °C), while accelerated in strength development, showed deterioration of their compressive strengths after 28 days when compared to samples cured at ambient or even slightly decreased temperatures (Rovnaník 2010).

Another experimental investigation found that 90 °C is the best curing temperature that produced the greatest unconfined compressive strengths in geopolymer. Moreover, temperatures higher than 90 °C resulted in drastically reduced the unconfined compressive strengths. Too high of a curing temperature results in rapid polycondensation, as well as excessive geopolymeric gel formation, which hinders unreacted alumina and silica dissolution. Additionally, excessively high temperatures cause pore solutions to rapidly evaporate, which can lead to incomplete geopolymerizations (Ahmari and Zhang 2012). Another similar study found higher curing temperatures (in this case, 60 °C) resulted in rapid strength development in the early curing stages of geopolymers—the first 28 days. After 28 days of curing, however, further strength development proved insignificant (Ridtirud et al. 2011).

According to Rovnaník (2010), high curing temperatures result in the development of large pores which affects the strength of geopolymers negatively. At temperatures of 80 and 60 °C, curing produces samples with high strength; however, with a reduction of strength after 4 weeks. On the other hand, geopolymers exposed to temperatures of 40 °C or 20 °C shows an increase in strength up to 4 weeks of testing (Rovnaník 2010). However, if the geopolymers were cured in water with say 20 °C, low strength would be obtained. This implication is believed to be as a result of leaching of dissolved species out from geopolymer's surfaces (Zuhua et al. 2009).

Also, high temperatures curing would certainly raise the tendency of geopolymer matrices to be cracked. The inclination to cracking in geopolymers is ascribed to the rapid loss of water and reduction of the open porosity (Perera et al. 2007). The rapid evaporation of mixing water eliminates the development of the desired strength. Therefore, it is often recommended to seal the samples of geopolymers at surfaces exposed during curing. A small amount of structural water should be held in the system to prevent cracking (Khalil and Merz 1994; Van Jaarsveld et al. 2002). Zuhua et al. (2009) pointed out that water traveling and liberated to the surface of the geopolymers through capillary action will as well result in the reduction of the structural water even in a sealed environment (Zuhua et al. 2009).

Geopolymer pastes pre-curing before being exposed to the regular curing has been proven to further augment its strength (Kani and Allahverdi 2009; Perera et al. 2007; Kim and Kim 2013). This pre-curing process is necessary for the consistent development of strength during the entire period, and good strength at initial phases can be obtained. Additionally, it lowers the porosity of geopolymers matrix, and more water can be retained within the paste (Perera et al. 2007). Kim and Kim suggested that the process of pre-curing at 75 °C for 3 h and then 4 weeks curing at room temperature produces high strength (51.06 MPa) metakaolin geopolymers (Kim and Kim 2013).

Nazari et al. (2011) examined the impact of curing temperature on the compressive strength of geopolymer produced from a combination of fly-ash and rice husk ash. A 24 h pre-curing time was conducted before casting to increase the consistency of the polymeric products before the heat was applied. Once the period of pre-curing was completed, the samples of geopolymers were exposed to temperatures ranging from 50 to 90 °C for 36 h. Based on the outcome from the inclusive developed strength, the maximum curing temperature of the entire mixtures at 1 and 4 weeks of curing was 80 °C. Additionally, the compressive strength of samples declined after increased curing temperatures and time of curing. This is for the reason that elevated curing temperatures would tear down the structural granular of the geopolymers. Elevated temperatures in curing also led to a contraction of the geopolymer gel and shrinkage of the matrices (Nazari et al. 2011). A method to reduce high temperatures curing time of high calcium fly-ash was proposed by Chindaprasirt et al. (2013). The outcome was that by exposing the specimens to microwave heating of 5 min combined with traditional oven treatment for 6 h at 60 °C, the compressive strength achieved was higher when in comparison to the specimens cured at for 24 h 60 °C with no microwave curing (Chindaprasirt et al. 2013).

1.2.5 Behaviour of Geopolymer Matrices at Elevated Temperatures

In recent years, investigations into the resistance of geopolymers concrete to high temperatures have been taken great interest, and many results with great potential have been obtained. One of the necessities for safety when designing construction structures is the aptitude to hold against high temperatures, which can result in spall due to the increased fragility and reduced permeability. The effect of high temperatures on the geopolymer produced using metakaolin and fly-ash in varied proportion of mixtures has been presented by Kong et al. (2007). The fly-ash-based product strength became firmer after it was exposed to relevant temperatures (800 °C). On the other hand, the strength of the subsequent metakaolin geopolymer went down after a similar exposure. The study revealed that the fly based geopolymers have more small pores in number enabling moisture escape during heating, therefore leading to the least damage to the geopolymers matrix. However, metakaolin geopolymers do not have similar structure. The strength development in fly-ash geopolymers during heating is also ascribed to the sintering responses of unreacted fly-ash particles (Kong et al. 2007).

Kong and Sanjayan (2010) in a different study carried out an investigation into the effect of high temperatures on geopolymer pastes made using fly-ash. Different experimental parameters such as aggregate sizing, aggregate type, specimen size and super plasticizer type have been examined. The study identified specimen size as well as aggregate size as two primary factors governing the geopolymers behavior at elevated temperatures (800 °C). According to Kong and Sanjayan (2010), the aggregate sizes with size greater than 10 mm produced greater performance in strength in the geopolymers concrete at increased temperatures. Furthermore, the reduction in the mechanical strength at increased temperatures was due to the thermal inconsistency between the aggregates and the geopolymer matrices (Kong and Sanjayan 2010).

An observation was made by Bakharev (2006) that the thermal stability of the geopolymers made with sodium activators was not high and considerable variations in the microstructure were observed. At temperatures of about 800 °C, the strength of the concrete went down as a result of increased average pore size where the amorphous structures were substituted with crystalline Na-feldspars. The opposite happened when the potassium silicate was chosen to be an activator as it remained mostly amorphous to 1200 °C. When firing geopolymer materials, the average pore size was declined, and improvements were observed on the compressive strength of geopolymer. In different study, the effect of fired temperatures on fly-ash-based geopolymers activated using sodium and potassium silicate has been reported. When exposing the material to increasing fired temperature ranging from 800 to 1200 °C, high shrinkage with increased changes in compressive strength were reported (Bakharev 2006).

1.2.6 *Fibre Reinforced Geopolymer Composites*

(i) *Synthetic Fibres Reinforced-Geopolymer Composites*

Mechanical performance of geopolymer matrices varies widely depending on the chemical composition and nature of aluminosilicate material, alkali activator and preparation process. However, geopolymers, as other ceramics, are generally brittle in nature (Pan and Sanjayan 2010; Pan et al. 2011). A composite material of geopolymer as a matrix and fibre or fabric as a reinforcement agent will extend the application of geopolymers in industry. Many researchers have studied the effect of several types of fibres on the mechanical properties of geopolymer composites (Lembo et al. 2014). For example, Puertas et al. (2003) studied the flexural and compressive strength of polypropylene fibre-reinforced mortars formed by adding slag, fly-ash, and slag/fly-ash at concentrations of 0, 0.5 and 1.0% by mortar volume. The researchers determined compressive strength at day 2 and/or day 28. Table 1.6 presents the results of flexural and compressive strength of all composites as tested at day 2 and day 28. The researchers' investigation found that adding polypropylene fibre at 0.5 and 1.0% concentration to slag-based geopolymer composites had no impact, at least no significant impact on compressive strength were measured at either day 2 or day 28. Additionally, they found that raising the concentration of polypropylene fibre in fly-ash-based geopolymer composite from 0%, to 0.5% and to 1.0% led to an increase in compressive strength by day 2 but an unexplained reduction in compressive strength by day 28. Slag/fly-ash-based geopolymer composites with polypropylene-fibre content increased from 0.5 to 1.0% were found to increase

Table 1.6 Flexural and compressive strengths of geopolymer composites as tested at day 2 and day 28 (Puertas et al. 2003)

Mortars	Fibre (%)	2 days		28 days	
		Flexural (MPa)	Compression (MPa)	Flexural (MPa)	Compression (MPa)
Slag	0	7.2	59.5	7.8	89.5
	0.5	6.5	60.1	7.6	90.0
	1	6.4	59.0	6.7	79.0
Fly-ash	0	3.9	24.5	6.8	39.4
	0.5	4.5	33.9	6.1	35.8
	1	4.9	34.3	5.5	26.9
Fly-ash/slag	0	3.6	11.8	4.6	30.0
	0.5	3.4	13.5	4.8	31.2
	1	2.9	13.6	4.8	30.1
Cement	0	6.3	39.1	7.8	53.0
	0.5	5.8	35.5	7.5	48.2
	1	5.4	38.9	7.4	47.6

compressive strength marginally at days 2 and 28 (Puertas et al. 2003). The study investigated the durability of the samples at 28 days; however, a further study on the effect of longer periods of times is necessary to gain a more useful information for the industrials sector.

In a study by Zhang et al. (2009), compressive strength in reinforced composite and non-reinforced composite were compared while using fly-ash and kaolin in geopolymer preparation. The report stated that after adding polypropylene-fibre of 0.5 wt%, the overall compressive strength increased. The increase was observed to be 67.8% on day 1 and at 19.5% on day 3. Although, when additional polypropylene at a greater proportion over 0.5 wt%, the rate of improvement in compressive strength decreased. The authors discovered a major improvement on the first and third day of test. Additional proportions equalling 0.25, 50 and 0.75 wt% were studied. When further fibres were added, the flexural strength of the composites improved (Zhang et al. 2009). The results are in contrast with the earlier results presented by Puertas et al. (2003) where the additional polypropylene fibres did not improve the composite's flexural strength on the 2nd and 28th test days. The failure in performance found in Puertas et al.'s study may be attributed to inferior workability because of added polypropylene fibres within the geopolymer paste (Shaikh 2013).

Yunsheng et al. (2006) studied the effect of increased amount of polyvinyl alcohol fibre on the impact strength of metakaolin-geopolymer composites. The strongest impact strength was found within the non-reinforced mortar at 450 N. The pure material was found to have a short internal displacement at 0.84 mm due to the brittleness and susceptible to failure after exceeding the peak load. The study reported a transformation from brittleness to ductility once polyvinyl alcohol fibres were increased to 2.0% by volume. The impact curve exceeded the peak load exhibiting a strain-hardening performance and showing ductility. The authors found that the peak load was 429.6 N and displacement was about 2.5 mm. When the peak load amount is exceeded, bearing capacity decreased until getting a displacement of 7.5 mm. According to Yunsheng et al. (2006) an additional 10 wt% of fly-ash resulted in the absorbed energy increased from 1833 to 2108 mJ (15% improvement). Although, greater concentrations above 30% presented a reduction in the impact resistance, and at 50 wt% the reduction begins to change significantly, stiffness and impact strength showed some enhancement. Yunsheng et al. (2006) reported that geopolymer with 50% fly-ash revealed an increase in toughness and stiffness, by 28.7% and 39.1%, respectively, and decreased in impact strength by 37.4% when compared to composites not using fly-ash (Yunsheng et al. 2006). As the researchers presented the optimum replacement amount of fly-ash; however, the ideal content of polyvinyl alcohol fibre has not been specified in the study. It seems to be critical to determine the optimum addition of fibres since the workability and physical properties of the pastes vary with varying the source materials, accordingly the ideal content of each type of fibres needed to design the composites with the highest mechanical properties will vary.

Natali et al. (2011) presented a research on the flexural strength in metakaolin/slag-geopolymer composites reinforced with E-glass, polyvinyl-alcohol (PVA), high tenacity carbon (HT) and polyvinyl-chloride (PVC). Improvements were observed

Table 1.7 Resilience and toughness indices for geopolymer composites (Natali et al. 2011)

Samples	Resilience (J/cm^2)	Toughness indices		
		I_5	I_{10}	I_{15}
GS	2.2	1.0	1.0	0
FeGS	3.1	4.6	6.5	8
FgGS	2.2	1.7	1.9	2.0
FpvaGS	2.6	3.1	3.4	3.5
FpvcGS	2.4	2.0	3.2	4.9

in the flexural strength when adding PVA, HT carbon, E-glass, and PVC. The study concluded that each fibre type has led to a satisfactory bridging effect. Additional 1.0 wt% of fibre to geopolymer composites, increased the flexural strength 30–70%, according to fibre type. The authors found that PVC and carbon fibres provided the best improvement in preventing post-crack behaviour (Natali et al. 2011). The reinforcements gave composites the greatest ductility at the first crack load. Table 1.7 pesents toughness indices for al geopolymer composites.

Lin et al. (2009) also presented a study on the flexural strength of carbon fibre reinforced geopolymers composites. A solution of K_2SiO_3 was used as an alkali activator and metakaolin as a raw aluminosilicate material. Composites were synthesized with changing amounts of carbon fibre (3.5, 4.5, 6 and 7.5 wt%). Table 1.8 presents density and mechanical results of all samples. The composite's density increased with increasing the fibre's content, this was due to the high density of carbon fibres. As the carbon-fibre volume fraction increased from 3.5 to 4.5%, flexural strength of composites also increased. At volume fraction 4.5%, flexural strength of geopolymer composite improved approximately by 475% if compared to the control sample. The addition of 6.0 and 7.5% of carbon fibres caused the strength to be reduced. Lin et al. (2009) argue that the reduction may be due to the damage in the fibre, in the shear stress formation in the fibre/matrix under high pressure forming (Lin et al. 2009).

Table 1.8 Density and mechanical results of all samples (Lin et al. 2009)

Materials	Density (g/cm^2)	Flexural strength (MPa)	Work of fracture (J/m^2)	Young's modulus (GPa)
Geopolymer	1.42	16.8 ± 0.7	54.2 ± 4.6	8.61 ± 0.43
3.5 vol.% Cf/geopolymer	1.42	91.3 ± 1.3	6435.3 ± 319.9	4.74 ± 0.63
4.5 vol.% Cf/geopolymer	1.49	96.6 ± 4.9	5915.2 ± 151.2	12.04 ± 0.45
6 vol.% Cf/geopolymer	1.56	87.4 ± 4.5	3926.3 ± 116.2	20.46 ± 1.61
7.5 vol.% Cf/geopolymer	1.67	42.0 ± 6.1	805.7 ± 49.9	17.77 ± 0.78

Li and Xu (2009) observed the impact strength, deformation and energy absorption in geopolymer matrices which have been reinforced using basalt fibres. The research suggested that basalt fibre-reinforced samples showed high strain rate of dependency. The results can be interpreted as strain rate increase contributed to the quality in impact properties. The basalt fibre reinforcement greatly enhanced energy absorption and deformation in concrete. The optimal loading for maxim energy absorption was suggested to be 0.3% volume fraction (Li and Xu 2009).

(ii) *Natural Fibres Reinforced-Geopolymer Composites*

Up to date, limited studies have investigated the effect of natural fibres on the mechanical performance of geopolymer composites. Alzeer and Mackenzie (2021) has recently provided an excellent review of geopolymer composites reinforced with natural fibres. Natural fibres have been shown to have the ability to resolve the problem of geopolymer brittleness and increase the ductility of geopolymer materials. In 2013, Alzeer and Mackenzie conducted a study on flexural reactions in metakaolin-based composites by adding 4.0–10 wt% content of unidirectional flax fibres. The composites showed improved flexural strength from 6.0 to 70 MPa after adding 10% of reinforcing fibre (Alzeer and Mackenzie 2013).

The same authors presented a study in the mechanical performance of geopolymer composites reinforced with marino wool and carpet wool. The wool fibre surface was given a chemical treatment to enhance reinforcing and alkali resistance properties. Table 1.9 shows the results of mechanical tests of all samples. The research found that unreinforced matrices exhibit brittle failure. However, the addition of the carpet wool that had been chemically-treated presented an increase in the averaged flexural strength by 50% and significant improvements in failure properties (Alzeer and Mackenzie 2012).

Table 1.9 Results of mechanical tests of geopolymer control sample and geopolymer composites reinforced with 5% of various types of wool fibres (Alzeer and Mackenzie 2012)

Fibre	Average fibre content (wt%)	Ultimate flexural strength (MPa)	Peak load (N)	Elastic modulus (GPa)
Merino wool	5	8.2 ± 1.5	25.4 ± 5.1	5.9 ± 0.5
Cleaned merino wool	5	9.1 ± 0.6	27.8 ± 2.3	10.2 ± 1.3
Treated merino wool	5	8.2 ± 0.8	25.0 ± 2.3	8.3 ± 1.1
Carpet wool	5	8.1 ± 2.3	19.8 ± 2.1	9.0 ± 1.9
Cleaned carpet wool	5	8.1 ± 1.5	25.8 ± 4.7	8.7 ± 2.0
Treated carpet wool	5	8.7 ± 0.5	26.6 ± 1.5	9.4 ± 0.6
Geopolymer matrix	0	5.8 ± 1.8	17.1 ± 6.0	9.6 ± 0.7

Alomayri et al. (2013b) presented their findings on mechanical and fracture performance of fly-ash geopolymer which has been reinforced using cotton fibres. The study showed that mechanical properties could be improved using cotton fibres in geopolymer composites. However, the study confirmed that increasing fibres contents caused a reduction of the composite's densities and an increase in porosities. Besides, agglomeration and poor dispersion of cotton fibres were observed when high content of the fibres were incorporated, causing a reduction on the mechanical strength of geopolymer composites. The researchers concluded that the flexural strength and fracture toughness optimum level occurred when fibre content used equaled to 0.5 wt% (Alomayri et al. 2013a; Alomayri and Low 2013). Figures 1.1 and 1.2 present the density and flexural strength of geopolymer composites.

The same authors suggested that to solve the problem of fibres agglomeration, geopolymer may reinforce with cotton fabric instead of cotton fibres. Therefore,

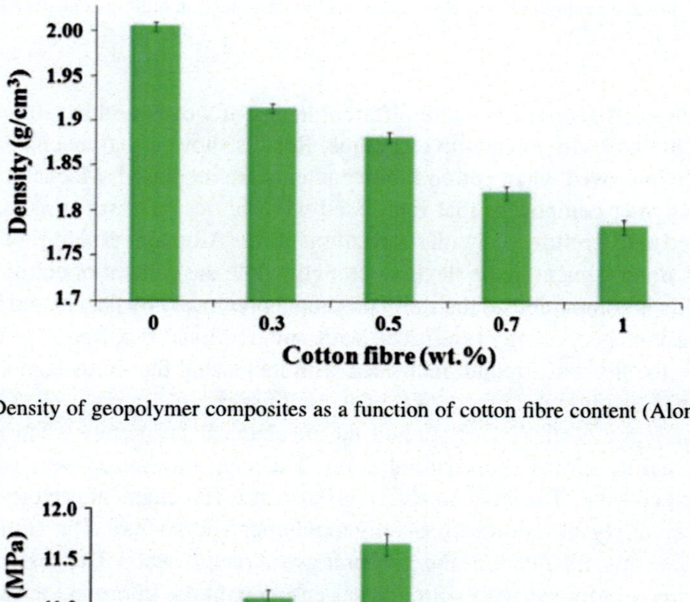

Fig. 1.1 Density of geopolymer composites as a function of cotton fibre content (Alomayri et al. 2013a)

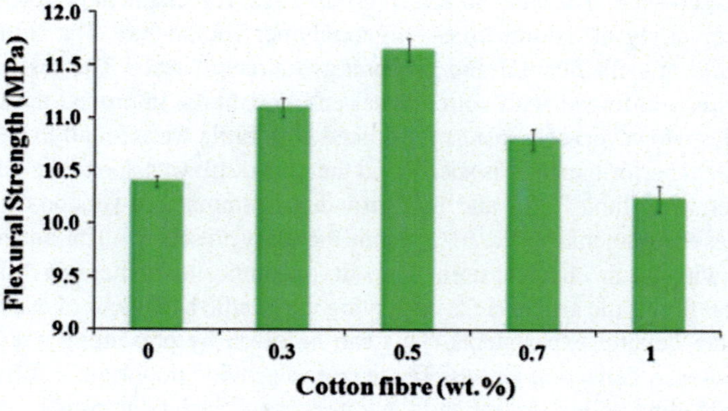

Fig. 1.2 Flexural strength of geopolymer composites as a function of cotton fibre content (Alomayri et al. 2013a)

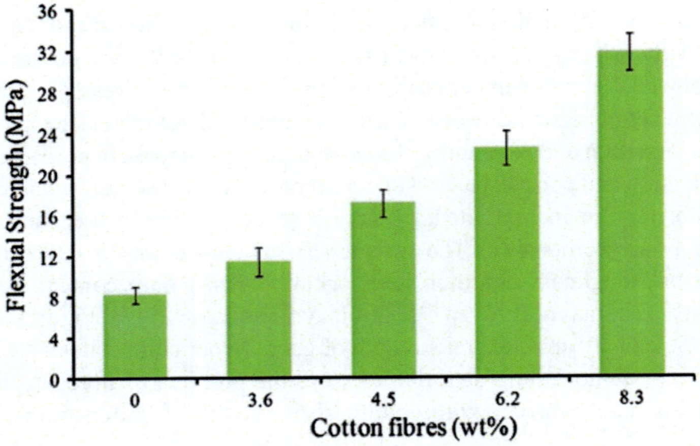

Fig. 1.3 Flexural strength of geopolymer composites as a function of fibre content (Alomayri et al. 2014)

they reinforced geopolymer with different layers of woven cotton fabrics (3.6, 4.5, 6.2 and 8.3 wt%) by the lay-up technique. Results showed that mechanical properties were improved when cotton fabric contents are increased. Mechanical strength of geopolymer composites that reinforced with fabrics gave superior results when compared to that reinforced with short cotton fibres (Alomayri et al. 2014). Figure 1.3 presents improvement in the flexural strength while the content of cotton increased. This study is comparable to the study presented previously by Alzeer and Mackenzie in the case of geopolymer reinforced with unidirectional flax fibres, in both investigations the flexural strength increased with increasing the fibres contents without reaching to an optimum amount of the natural fibres.

Korniejenko et al. (2016) studied the mechanical properties when reinforcing geopolymer by adding short natural fibres. The short fibres used were cotton, sisal, raffia and coconut. The study analysed influences on mechanical performance when adding a variety of natural fibres to geopolymer composites. The study included the results of both flexural and compressive strength tests. The test proved that composites reinforced with cotton, sisal and coir fibres improved the mechanical properties. However, composites reinforced with raffia were found to be poor in its mechanical performance. It is speculated that this result was due to the fibre size and characteristics. Tables 1.10 and 1.11 provides a summarized version of the study's outcomes (Korniejenko et al. 2016). While the study presented an interesting comparison showing the mechanical strength results of composites containing different types of natural fibres, the article lacks displaying the ductile behaviour of each composite during the flexural experiments. This can be given by presenting stress–strain or load-mid-span deflection curves. Those curves provide important information such as peak-loading points, strains and the toughness of each composite.

Table 1.10 Compressive strength values of the natural fibres-reinforced geopolymer composites at 28 days (Korniejenko et al. 2016)

Sample	MPa	Standard deviation of recorded values of strengths
Geopolymer (matrix)	24.78	1.89
Geopolymer with coir fibers (1%)	31.36	10.10
Geopolymer with cotton fibers (1%)	28.42	5.30
Geopolymer with raffia fibers (1%)	13.66	1.71
Geopolymer with sisal fibers (1%)	25.16	3.43

Table 1.11 Flexural strength values of the natural fibres-reinforced geopolymer composites at 28 days (Korniejenko et al. 2016)

Sample	MPa	Standard deviation of recorded values of strengths
Geopolymer (matrix)	5.55	0.72
Geopolymer with coir fibers (1%)	5.25	0.57
Geopolymer with cotton fibers (1%)	5.85	0.78
Geopolymer with raffia fibers (1%)	3.05	0.35
Geopolymer with sisal fibers (1%)	5.90	0.14

Al Bakri et al. (2013) provided a study in the compressive strength of fly-ash-geopolymer which have been reinforced using wood fibres. To prepare, the fly-ash was activated using 2.5 ratio of Na_2SiO_3:NaOH. Short-wood fibres were then introduced into the compound using percentages ranging between 10 and 50 wt%. The results showed that an increase of wood fibre content caused a decrease in compressive strength of the composites at the 7th and 14th day. It is theorized that the result is due to wood-fibre reacting as a filler in the matrix. Reduction is due to an increased surface area of filler materials that bonds to the geopolymer matrix which created a decrease in geopolymer's surface area (Al Bakri et al. 2013). The study lacks a control geopolymer sample for comparison; additionally, it looks to be a necessity to investigate the compressive strength of composites containing fibres less than 10 wt% in order to find an optimum point with the maximum compressive strength.

Chen et al. (2014) observed the flexural strength of sweet sorghum reinforced geopolymer composites. The research found that the strength increased with fibre content up to 2.0 wt%. However, when fibre exceeds 2.0% the toughness, tensile and flexural strength decreased. Research concluded that 2.0 wt% in sweet sorghum fibre is the effective amount which allows stronger loads to be distributed throughout the composite. The increase delayed micro-crack growth and increasing flexural strength. However, going beyond 2.0% caused poor workability, and a disproportional dispersion of the fibres causing air bubbles to become entrapped. The weakness caused concentrations of stress and a reduction of the composite's flexural strength (da Silva Alves et al. 2019; Chen et al. 2014).

Correia et al. (2013) provided an investigation on the compressive and impact properties of geopolymer composites reinforced using plant fibres. The researchers used 3.0% volume fraction of pineapple leaf fibre (PALF) and sisal fibre to the metakaolin-based composites. The findings showed that pineapple leaf fibre and sisal fibres used as reinforcements created an improvement to the compressive and impact strength results. The strength of the geopolymer composites with added sisal was observed to be higher than pineapple leaf fibre. The indications of the study showed that a composite which uses a natural fibre that can be observed to increase mechanical properties should be proposed (Poletanovic et al. 2020; Silva et al. 2020; Ferreira et al. 2019; Correia et al. 2013).

(iii) *Durability of Natural Fibre Reinforced Composites*

Durability of natural fibre in composites is the capability to resist the degradation processes either by external damage such as chloride and alkali attack or internal damage such as compatibility between fibres and the resin matrices (Mohr et al. 2005; Walker et al. 2014; Gram 1983; Santos et al. 2015; Wei and Meyer 2015). Geopolymers are generally high alkali matrices. Therefore, there are a concern in incorporating plant fibres in alkali-based pastes. The main concern is regarding the long-term durability of natural fibre reinforced composites. Natural fibres can be degraded and damaged in high-alkaline environment; thereby adversely influencing the mechanical performance and durability of the composites (Hakamy et al. 2016; Aly et al. 2011; Yan and Chouw 2015). Natural fibre degradations in alkaline environments was studied in early 1980s (Gram 1983). The degradation mechanism was described as the decomposition of hemicellulose and linen which leads to the splitting of natural fibres into micro-fibrils. This effect has been observed using SEM in the case of jute fibres in cement matrix, where the natural fibres split-up and fibrillised resulting in reduction in the tensile strength of jute fibres by 76% (Velpari et al. 1980). To reduce the degradation impact of natural fibres in alkaline matrices, nanoparticles can play an important role. The effect of nanoclay particles on the durability of flax fibres reinforced cement composites at 28 days and after 50 wet/dry cycles has been examined (Aly et al. 2011). Based on SEM examination, samples loaded with 2.5 wt% nanoclay particles were observed to show lower deterioration in the flexural strength when compared to its counterpart control samples. This was attributed to the effect of nanoclay particles in decreasing the degradation of flax fibres. However, there is no known published research discussing the durability of geopolymer composites reinforced with natural fibre, or the deterioration of cellulose fibres in geopolymer matrices.

(iv) *Nanofiller Reinforced-Geopolymer Composites*

The advances in the field of nanotechnology have in recent years been exploited in the development and manipulation of the properties of geopolymer in the form of concrete, mortar and paste (Assaedi et al. 2019; Jindal and Sharma 2021).

Nanotechnology is the technique of manipulating nanomaterials, having the size in the ranges of 1–100 nm and possessing a higher surface area to volume ratio. The higher specific surface area of nanoparticles increases their reactivity and accelerates the chemical reaction (Lazaro et al. 2016). Nanoparticles have been found to modify the microstructure of the concrete and mortar matrix at the atomic level and further improves the properties of concrete in fresh as well as in the hardened state. An excellent review on the uses of various nanomaterials in geopolymer composites has been described by Jindal and Sharma (2021). In this review, the authors discuss the effects of various types of nanomaterials such as nano-silica, nano-clay, nano-alumina, carbon nanotubes, and nano-TiO_2 on the workability, setting times, geopolymerisation, strength properties, and durability properties of geopolymer composites.

1.2.7 *Applications of Geopolymers*

While discussion of geopolymer application thus far has concentrated on the construction industry and overwhelmingly on its use as a concrete, there are a variety of other roles for this material, including use in the aerospace, civil engineering, non-ferrous foundries, motor vehicle and plastic industries (Davidovits 1999). The potential for application is categorised by Davidovits, who matches application type to the Si:Al ratio. The range of applications includes fire or heat resistant materials, sealants for construction industry, tooling for aeronautics, foundry equipments, low CO_2 cements and concretes as well as encapsulation of radioactive substances and toxic wastes (Davidovits 1999).

An interesting sustainable development application of geopolymer materials is in the management of toxic and hazardous waste, because geopolymer materials have properties similar to zeolitic materials, which are known for their capacity to absorb toxic and hazardous waste types (Davidovits 2002).

Balaguru et al. (1997) investigated the use of geopolymers for the purposes of fastening carbon fabric to the surface of reinforced concrete beams. They find that geopolymer offers excellent adhesion of the carbon fabric's inter-laminar to the reinforced concrete surface. They report that the geopolymer is chemically compatible with the reinforced concrete, that it does not undergo UV radiation degradation, and that it is fire resistant.

In Australia to date there has been broad application of geopolymer technology, with notable developments in building products such as chemically-resistant wall panels and fire-resistant construction materials; further afield, geopolymer technology has been applied in railway sleepers, masonry units, protective coatings, high-performing fibre-reinforced laminates, repairs materials and sewer pipeline products (Gourley 2003; Gourley and Johnson 2005).

References

Ahmari S, Zhang L (2012) Production of eco-friendly bricks from copper mine tailings through geopolymerization. Constr Build Mater 29:323–331

Ahmari S, Ren X, Toufigh V, Zhang L (2012) Production of geopolymeric binder from blended waste concrete powder and fly ash. Constr Build Mater 35:718–729

Al Bakri M, Mohd A, Izzat AM, Muhammad Faheem M, Kamarudin H, Khairul Nizar I, Bnhussain M, Rafiza A, Zarina Y, Liyana J (2013) Feasibility of producing wood fibre-reinforced geopolymer composites (WFRGC). Adv Mater Res 626:918–925

Alomayri T, Low IM (2013) Synthesis and characterization of mechanical properties in cotton fiber-reinforced geopolymer composites. J Asian Ceram Soc 1:30–34

Alomayri T, Shaikh FUA, Low IM (2013a) Characterisation of cotton fibre-reinforced geopolymer composites. Compos B Eng 50:1–6

Alomayri T, Shaikh FUA, Low IM (2013b) Thermal and mechanical properties of cotton fabric-reinforced geopolymer composites. J Mater Sci 48:6746–6752

Alomayri T, Shaikh FUA, Low IM (2014) Synthesis and mechanical properties of cotton fabric reinforced geopolymer composites. Compos B Eng 60:36–42

Alonso S, Palomo A (2001) Alkaline activation of metakaolin and calcium hydroxide mixtures: influence of temperature, activator concentration and solids ratio. Mater Lett 47:55–62

Aly M, Hashmi MSJ, Olabi AG, Messeiry M, Hussain AI, Abadir EF (2011) Effect of nano-clay and waste glass powder on the properties of flax fibre reinforced mortar. J Eng Appl Sci 6:19–28

Alzeer M, Mackenzie KD (2012) Synthesis and mechanical properties of new fibre-reinforced composites of inorganic polymers with natural wool fibres. J Mater Sci 47:6958–6965

Alzeer M, Mackenzie K (2013) Synthesis and mechanical properties of novel composites of inorganic polymers (geopolymers) with unidirectional natural flax fibres (phormium tenax). Appl Clay Sci 75–76:148–152

Alzeer M, Mackenzie K (2021) Composites of inorganic polymers (geopolymers) reinforced with natural fibres. In: Low IM, Dong Y (eds) Composite materials: manufacturing, properties and applications, Chap 5. Elsevier

Andre A (2006) Fibres for Strengthening of Timber Structures. Ph.D. Thesis, Department of Civil and Environmental Engineering, Lulea University, Sweden

Anuar H, Zuraida A (2011) Improvement in mechanical properties of reinforced thermoplastic elastomer composite with kenaf bast fibre. Compos Part B: Eng 42:462–465

Arioz E, Arioz O, Koçkar OM (2013) The effect of curing conditions on the properties of geopolymer samples. Int J Chem Eng Appl 4:423–426

Assaedi H, Alomayri T, Shaikh F, Low IM (2019) Influence of nano silica particles on durability of flax fabric reinforced geopolymer composites. Materials 12. https://doi.org/10.3390/ma12091459

Aydin S, Baradan B (2012) Mechanical and microstructural properties of heat cured alkali-activated slag mortars. Mater Des 35:374–383

Bakharev T (2005) Geopolymeric materials prepared using class F fly ash and elevated temperature curing. Cem Concr Res 35:1224–1232

Bakharev T (2006) Thermal behaviour of geopolymers prepared using class F fly ash and elevated temperature curing. Cem Concr Res 36:1134–1147

Balaguru P, Kurtz S, Rudolph J (1997) Geopolymer for repair and rehabilitation of reinforced concrete beams. Institut Géopolymère, Saint-Quentin

Barbosa VFF, Mackenzie KJD, Thaumaturgo C (2000) Synthesis and characterisation of materials based on inorganic polymers of alumina and silica: sodium polysialate polymers. Int J Inorg Mater 2:309–317

Bax B, Müssig J (2008) Impact and tensile properties of PLA/cordenka and PLA/flax composites. Compos Sci Technol 68:1601–1607

Beckermann G (2007) Performance of hemp-fibre reinforced polypropylene composite materials. PhD thesis. The University of Waikato

Bhat GS, Kamath MG, Mueller D, Parikh DV, Mclean M (2004) Cotton-based composites for automotive applications. GPEC Conference Proceedings

Bohlooli H, Nazari A, Khalaj G, Kaykha MM, Riahi S (2012) Experimental investigations and fuzzy logic modeling of compressive strength of geopolymers with seeded fly ash and rice husk bark ash. Compos B Eng 43:1293–1301

Chaudhary V, Gohil P (2013) Stress analysis in cotton polyester composite material. Int J Metall Mater 3:15–22

Chen R, Ahmari S, Zhang L (2014) Utilization of sweet sorghum fiber to reinforce fly ash-based geopolymer. J Mater Sci 49:2548–2558

Cheng T, Chiu J (2003) Fire-resistant geopolymer produced by granulated blast furnace slag. Miner Eng 16:205–210

Chen-Tan NW, Van Riessen A, Ly CV, Southam DC (2009) Determining the reactivity of a fly ash for production of geopolymer. J Am Ceram Soc 92:881–887

Chindaprasirt P, Chareerat T, Sirivivatnanon V (2007) Workability and strength of coarse high calcium fly ash geopolymer. Cem Concr Compos 29:224–229

Chindaprasirt P, Jaturapitakkul C, Chalee W, Rattanasak U (2009) Comparative study on the characteristics of fly ash and bottom ash geopolymers. Waste Manage 29:539–543

Chindaprasirt P, Rattanasak U, Taebuanhuad S (2013) Resistance to acid and sulfate solutions of microwave-assisted high calcium fly ash geopolymer. Mater Struct 46:375–381

Cioffi R, Maffucci L, Santoro L (2003) Optimization of geopolymer synthesis by calcination and polycondensation of a kaolinitic residue. Resour Conserv Recycl 40:27–38

Correia EAS, Torres SM, De Oliveira Alexandre ME, Gomes KC, Barbosa N, De Barros SR (2013) Mechanical performance of natural fibers reinforced geopolymer composites. Mater Sci Forum. Trans Tech Publications

da Silva Alves LC, dos Reis Ferreira RA, Bellini Machado L, de Castro Motta LA (2019) Optimization of metakaolin-based geopolymer reinforced with sisal fibers using response surface methology. Ind Crops Prod 139:111551

Davidovits J (1991) Years of successes and failures in geopolymer applications. Market trends and potential breakthroughs. In: Keynote conference on geopolymer conference

Davidovits J (1994a) Global warming impact on the cement and aggregates industries. World Resour Rev 6:263–278

Davidovits J (1994b) Properties of geopolymer cements. In: Proceedings first international conference, Saint-Quentin

Davidovits J (1999) Chemistry of geopolymeric systems, terminology. In: Proceedings of geopolymer '99 international conference. Institut Géopolymère, Saint-Quentin

Davidovits J (2002) Geopolymers—inorganic polymeric new materials. J Therm Anal 37:1633–1656

Davidovits J (2008) Geopolymer chemistry and applications. Geopolymer Institute, Saint-Quentin

De Vargas AS, Dal Molin DCC, Vilela ACF, Silva FJD, Pavão B, Veit H (2011) The effects of Na_2O/SiO_2 molar ratio, curing temperature and age on compressive strength, morphology and microstructure of alkali-activated fly ash-based geopolymers. Cem Concr Compos 33:653–660

Detphan S, Chindaprasirt P (2009) Preparation of fly ash and rice husk ash geopolymer. Int J Miner Metall Mater 16:720–726

Dhakal H, Zhang Z, Richardson M (2007) Effect of water absorption on the mechanical properties of hemp fibre reinforced unsaturated polyester composites. Compos Sci Technol 67:1674–1683

Dimas D, Giannopoulou I, Panias D (2009) Polymerization in sodium silicate solutions: a fundamental process in geopolymerization technology. J Mater Sci 44:3719–3730

Duxson P, Fernández-Jiménez A, Provis JL, Lukey GC, Palomo A, Van Deventer JSJ (2007a) Geopolymer technology: the current state of the art. J Mater Sci 42:2917–2933

Duxson P, Mallicoat SW, Lukey GC, Kriven WM, Van Deventer JSJ (2007b) The effect of alkali and Si/Al ratio on the development of mechanical properties of metakaolin-based geopolymers. Colloids Surf A Physicochem Eng Asp 292:8–20

Fernández-Jiménez A, Palomo A, Criado M (2005) Microstructure development of alkali-activated fly ash cement: a descriptive model. Cem Concr Res 35:1204–1209

Ferone C, Colangelo F, Roviello G, Asprone D, Menna C, Balsamo A, Prota A, Cioffi R, Manfredi G (2013) Application-oriented chemical optimization of a metakaolin based geopolymer. Materials 6:1920–1939

Ferreira VF, Pinheiro FI, de Souza FS, Mei HIL, Lona MFL (2019) Polymer composites reinforced with natural fibers and nanocellulose in the automotive industry: a short review. J Compos Sci 3. https://doi.org/10.3390/jcs3020051

Fervel V, Mischler S, Landolt D (2003) Lubricating properties of cotton transfer films studied with a pin-on-disc apparatus. Wear 254:492–500

Gao K, Lin KL, Wang D, Hwang CL, Shiu HS, Chang YM, Cheng TW (2014) Effects SiO_2/Na_2O molar ratio on mechanical properties and the microstructure of nano-SiO_2 metakaolin-based geopolymers. Constr Build Mater 53:503–510

Giasuddin HM, Sanjayan JG, Ranjith PG (2013) Strength of geopolymer cured in saline water in ambient conditions. Fuel 107:34–39

Gorhan G, Kurklu G (2014) The influence of the NaOH solution on the properties of the fly ash-based geopolymer mortar cured at different temperatures. Compos B Eng 58:371–377

Gourley J (2003) Geopolymers; opportunities for environmentally friendly construction materials. In: Materials conference: adaptive materials for a modern society. Institute of Materials Engineering Australia, Sydney

Gourley JT, Johnson GB (2005) Developments in geopolymer precast concrete. In: International workshop on geopolymers and geopolymer concrete, Perth

Gram H-E (1983) Methods for reducing the tendency towards embrittlement in sisal fibre concrete. Nord Concr Res 5:62–71

Gunasekara M, Law D, Setunge S (2014) Effect of composition of fly ash on compressive strength of fly ash based geopolymer mortar. In: 23rd Australasian conference on the mechanics of structures and materials (ACMSM23). Southern Cross University, Lismore

Guo X, Shi H, Dick WA (2010) Compressive strength and microstructural characteristics of class C fly ash geopolymer. Cem Concr Compos 32:142–147

Hakamy A, Shaikh FUA, Low IM (2016) Effect of calcined nanoclay on the durability of NaOH treated hemp fabric-reinforced cement nanocomposites. Mater Des 92:659–666

Hanjitsuwan S, Hunpratub S, Thongbai P, Maensiri S, Sata V, Chindaprasirt P (2014) Effects of NaOH concentrations on physical and electrical properties of high calcium fly ash geopolymer paste. Cem Concr Compos 45:9–14

Hashmi SAR, Dwivedi UK, Chand N (2006) Friction and sliding wear study of UHMWPE modified cotton fibre polyester composites. Tribol Lett 21:79–87

Hashmi SAR, Dwivedi UK, Chand N (2007) Graphite modified cotton fibre reinforced polyester composites under sliding wear conditions. Wear 262:1426–1432

He J, Zhang G, Zhang J, Yu Y (2012) The strength and microstructure of two geopolymers derived from metakaolin and red mud-fly ash admixture: a comparative study. Constr Build Mater 30:80–91

He J, Jie Y, Zhang J, Yu Y, Zhang G (2013) Synthesis and characterization of red mud and rice husk ash-based geopolymer composites. Cem Concr Compos 37:108–118

Heah C, Kamarudin H, Al Bakri AM, Binhussain M, Luqman M, Nizar IK, Ruzaidi C, Liew Y (2011) Effect of curing profile on kaolin-based geopolymers. Phys Procedia 22:305–311

Hendra P, Peer Mohamed A (2010) A study of cotton–ramie fabric reinforced composites. Text Technol, 1–12

Hu M, Zhu X, Long F (2009) Alkali-activated fly ash-based geopolymers with zeolite or bentonite as additives. Cem Concr Compos 31:762–768

Islam A, Alengaram UJ, Jumaat MZ, Bashar II (2014) The development of compressive strength of ground granulated blast furnace slag-palm oil fuel ash-fly ash based geopolymer mortar. Mater Des 56:833–841

Jansen MS, Christiansen MU (2015) Effect of water-solids ratio on the compressive strength and morphology of fly ash-waste glass geopolymer mortars. In: WOCA (ed) 2015 world of coal ash (WOCA), Nashville

Jindal BB, Sharma S (2021) Geopolyymer composites modified with nanomaterials. In: Low IM, Dong Y (eds) Composite materials: manufacturing, properties and applications, Chap 2. Elsevier

Jun Y, Oh J (2015) Use of gypsum as a preventive measure for strength deterioration during curing in class F fly ash geopolymer system. Materials 8:3053–3067

Kani EN, Allahverdi A (2009) Effects of curing time and temperature on strength development of inorganic polymeric binder based on natural pozzolan. J Mater Sci 44:3088–3097

Karus M, Ortmann S, Vogt D (2005) Natural fibre in automotive interior: all natural on the inside? Kunstst Plast Eur 7:1–3

Khale D, Chaudhary R (2007) Mechanism of geopolymerization and factors influencing its development: a review. J Mater Sci 42:729–746

Khalil MY, Merz E (1994) Immobilization of intermediate-level wastes in geopolymers. J Nucl Mater 211:141–148

Kim H, Kim Y (2013) Relationship between compressive strength of geo-polymers and pre-curing conditions. Appl Microsc 43:155–163

Komljenović M, Baščarević Z, Bradić V (2010) Mechanical and microstructural properties of alkali-activated fly ash geopolymers. J Hazard Mater 181:35–42

Komnitsas K, Zaharaki D (2007) Geopolymerisation: a review and prospects for the minerals industry. Miner Eng 20:1261–1277

Kong DLY, Sanjayan JG (2010) Effect of elevated temperatures on geopolymer paste, mortar and concrete. Cem Concr Res 40:334–339

Kong DLY, Sanjayan JG, Sagoe-Crentsil K (2007) Comparative performance of geopolymers made with metakaolin and fly ash after exposure to elevated temperatures. Cem Concr Res 37:1583–1589

Kong DLY, Sanjayan JG, Sagoe-Crentsil K (2008) Factors affecting the performance of metakaolin geopolymers exposed to elevated temperatures. J Mater Sci 43:824–831

Korniejenko K, Frączek E, Pytlak E, Adamski M (2016) Mechanical properties of geopolymer composites reinforced with natural fibers. Procedia Eng 151:388–393

Kriven WM, Bell JL, Gordon M (2003) Microstructure and microchemistry of fully-reacted geopolymers and geopolymer matrix composites. Ceram Trans 153:227–250

Kutchko BG, Kim AG (2006) Fly ash characterization by SEM–EDS. Fuel 85:2537–2544

Latella BA, Perera DS, Durce D, Mehrtens EG, Davis J (2008) Mechanical properties of metakaolin-based geopolymers with molar ratios of Si/Al $\approx$ 2 and Na/Al $\approx$ 1. J Mater Sci 43:2693–2699

Law D, Adam A, Molyneaux T, Patnaikuni I, Wardhono A (2014) Long term durability properties of class F fly ash geopolymer concrete. Mater Struct 48:1–11

Lazaro A, Yu QL, Brouwers HJH (2016) Nanotechnologies for sustainable construction. Sustain Constr Mater. https://doi.org/10.1016/b978-0-08-100370-1.00004-4

Lee WKW, Van Deventer JSJ (2002) Structural reorganisation of class F fly ash in alkaline silicate solutions. Colloids Surf A Physicochem Eng Asp 211:49–66

Lembo K, Lokuge W, Karunasena W (2014) Geopolymer concrete with FRP confinement. In: 23rd Australasian conference on the mechanics of structures and materials (ACMSM23). Southern Cross University, Lismore

Li G, Yu Y, Zhao Z, Li J, Li C (2003) Properties study of cotton stalk fiber/gypsum composite. Cem Concr Res 33:43–46

Li W, Xu J (2009) Impact characterization of basalt fiber reinforced geopolymeric concrete using a 100-mm-diameter split Hopkinson pressure bar. Mater Sci Eng A 513:145–153

Li Q, Xu H, Li F, Li P, Shen L, Zhai J (2012) Synthesis of geopolymer composites from blends of CFBC fly and bottom ashes. Fuel 97:366–372

Liefke E (1999) Industrial applications of foamed inorganic polymers. In: Geopolymer '99 international conference proceedings

Liew YM, Heah CY, Mohd Mustafa AB, Kamarudin H (2016) Structure and properties of clay-based geopolymer cements: a review. Prog Mater Sci 83:595–629

Lin T, Jia D, Wang M, He P, Liang D (2009) Effects of fibre content on mechanical properties and fracture behaviour of short carbon fibre reinforced geopolymer matrix composites. Bull Mater Sci 32:77–81

Lizcano M, Gonzalez A, Basu S, Lozano K, Radovic M (2012) Effects of water content and chemical composition on structural properties of alkaline activated metakaolin-based geopolymers. J Am Ceram Soc 95:2169–2177

Mallick PK (2007) Fibre-reinforced composites: materials, manufacturing, and design. CRC/Taylor & Francis, Boca Raton

Mishra A, Choudhary D, Jain N, Kumar M, Sharda N, Dutt D (2008) Effect of concentration of alkaline liquid and curing time on strength and water absorption of geopolymer concrete. J Eng Appl Sci 3:14–18

Mohr BJ, Nanko H, Kurtis KE (2005) Durability of kraft pulp fiber–cement composites to wet/dry cycling. Cem Concr Compos 27:435–448

Monteiro S, Lopes F, Ferreira A, Nascimento D (2009) Naturalfiber polymer-matrix composites: cheaper, tougher, and environmentally friendly. JOM J Miner, Metals Mater Soc 61:17–22

Müssig J (2008) Cotton fibre-reinforced thermosets versus ramie composites: a comparative study using petrochemical-and agro-based resins. J Polym Environ 16:94–102

Natali A, Manzi S, Bignozzi M (2011) Novel fiber-reinforced composite materials based on sustainable geopolymer matrix. Procedia Eng 21:1124–1131

Nazari A, Bagheri A, Riahi S (2011) Properties of geopolymer with seeded fly ash and rice husk bark ash. Mater Sci Eng A 528:7395–7401

Palomo A, Blanco-Varela MT, Granizo ML, Puertas F, Vazquez T, Grutzeck MW (1999a) Chemical stability of cementitious materials based on metakaolin. Cem Concr Res 29:997–1004

Palomo A, Grutzeck MW, Blanco MT (1999b) Alkali-activated fly ashes: a cement for the future. Cem Concr Res 29:1323–1329

Pan Z, Sanjayan JG (2010) Stress–strain behaviour and abrupt loss of stiffness of geopolymer at elevated temperatures. Cem Concr Compos 32:657–664

Pan Z, Sanjayan JG, Rangan BV (2011) Fracture properties of geopolymer paste and concrete. Mag Concr Res 63:763–771

Pangdaeng S, Sata V, Aguiar JB, Pacheco-Torgal F, Chindaprasirt P (2015) Apatite formation on calcined kaolin-white Portland cement geopolymer. Mater Sci Eng C 51:1–6

Part WK, Ramli M, Cheah CB (2015) An overview on the influence of various factors on the properties of geopolymer concrete derived from industrial by-products. Constr Build Mater 77:370–395

Perera DS, Vance ER, Finnie KS, Blackford MG, Hanna JV, Cassidy DJ (2005) Disposition of water in metakaolinite based geopolymers. Adv Ceram Matrix Compos 175:225–236

Perera DS, Uchida O, Vance ER, Finnie KS (2007) Influence of curing schedule on the integrity of geopolymers. J Mater Sci 42:3099–3106

Perná I, Hanzlíček T, Šupová M (2014) The identification of geopolymer affinity in specific cases of clay materials. Appl Clay Sci 102:213–219

Phair JW, Van Deventer JSJ (2002) Effect of the silicate activator pH on the microstructural characteristics of waste-based geopolymers. Int J Miner Process 66:121–143

Poletanovic B, Dragas J, Ignjatovic I, Komljenovic M, Merta I (2020) Physical and mechanical properties of hemp fibre reinforced alkali-activated fly ash and fly ash/slag mortars. Constr Build Mater 259:119677

Puertas F, Amat T, Fernández-Jiménez A, Vázquez T (2003) Mechanical and durable behaviour of alkaline cement mortars reinforced with polypropylene fibres. Cem Concr Res 33:2031–2036

Raftoyiannis IG (2012) Experimental testing of composite panels reinforced with cotton fibers. J Compos Mater 2:31–39

Rahier H, Simons W, Van Mele B, Biesemans M (1997) Low-temperature synthesized aluminosilicate glasses: part III Influence of the composition of the silicate solution on production, structure and properties. J Mater Sci 32:2237–2247

Rahier H, Wastiels J, Biesemans M, Willlem R, Van Assche G, Van Mele B (2007) Reaction mechanism, kinetics and high temperature transformations of geopolymers. J Mater Sci 42:2982–2996

Rattanasak U, Chindaprasirt P (2009) Influence of NaOH solution on the synthesis of fly ash geopolymer. Miner Eng 22:1073–1078

Rattanasak U, Chindaprasirt P, Suwanvitaya P (2010) Development of high volume rice husk ash alumino silicate composites. Int J Miner Metall Mater 17:654–659

Rees CA, Provis JL, Lukey GC, Van Deventer JSJ (2008) The mechanism of geopolymer gel formation investigated through seeded nucleation. Colloids Surf A 318:97–105

Rickard WDA, Williams R, Temuujin J, Van Riessen A (2011) Assessing the suitability of three Australian fly ashes as an aluminosilicate source for geopolymers in high temperature applications. Mater Sci Eng A 528:3390–3397

Rickard WDA, Kealley CS, Van Riessen A, Biernaki J (2015) Thermally induced microstructural changes in fly ash geopolymers: experimental results and proposed model. J Am Ceram Soc 98:929–939

Ridtirud C, Chindaprasirt P, Pimraksa K (2011) Factors affecting the shrinkage of fly ash geopolymers. Int J Miner Metall Mater 18:100–104

Rovnaník P (2010) Effect of curing temperature on the development of hard structure of metakaolin-based geopolymer. Constr Build Mater 24:1176–1183

Rowles M, O'Connor B (2003) Chemical optimisation of the compressive strength of aluminosilicate geopolymers synthesised by sodium silicate activation of metakaolinite. J Mater Chem 13:1161–1165

Rukmini K, Ramaraj B, Shetty SK, Taraiya A, Bandyopadhyay S (2013) Development of ecofriendly cotton fabric reinforced polypropylene composites: mechanical, thermal, and morphological properties. Adv Polym Technol 32:1–9

Salih MA, Abang Ali AA, Farzadnia N (2014) Characterization of mechanical and microstructural properties of palm oil fuel ash geopolymer cement paste. Constr Build Mater 65:592–603

Santos SFD, Tonoli GHD, Mejia J, Fiorelli J, Savastano H Jr (2015) Non-conventional cement-based composites reinforced with vegetable fibers: a review of strategies to improve durability. Mater Constr 65:041

Sathonsaowaphak A, Chindaprasirt P, Pimraksa K (2009) Workability and strength of lignite bottom ash geopolymer mortar. J Hazard Mater 168:44–50

Sgriccia N, Hawley MC, Misra M (2008) Characterization of natural fiber surfaces and natural fiber composites. Compos Part A: Appl Sci Manuf 39:1632–1637

Shaikh FUA (2013) Review of mechanical properties of short fibre reinforced geopolymer composites. Constr Build Mater 43:37–49

Silva G, Kim S, Aguilar R, Nakamatsu J (2020) Natural fibers as reinforcement additives for geopolymers—a review of potential eco-friendly applications to the construction industry. Sustain Mater Technol 23:e00132. https://doi.org/10.1016/j.susmat.2019.e00132

Soleimani M, Naghizadeh R, Mirhabibi A, Golestanifard F (2012) Effect of calcination tmperature of the kaolin and molar Na_2O/SiO_2 activator ratio on physical and microstructural properties of metalkaolinbased geopolymers. Iran J Mater Sci Eng 9:43–51

Somna K, Jaturapitakkul C, Kajitvichyanukul P, Chindaprasirt P (2011) NaOH-activated ground fly ash geopolymer cured at ambient temperature. Fuel 90:2118–2124

Sreenivasan VS, Ravindran D, Manikandan V, Narayanasamy R (2011) Mechanical properties of randomly oriented short sansevieria cylindrica fibre/polyester composites. Mater Des 32:2444–2455

Sukmak P, Horpibulsuk S, Shen SL (2013) Strength development in clay-fly ash geopolymer. Constr Build Mater 40:566–574

Tchakoute Kouamo H, Mbey JA, Elimbi A, Kenne Diffo BB, Njopwouo D (2013) Synthesis of volcanic ash-based geopolymer mortars by fusion method: effects of adding metakaolin to fused volcanic ash. Ceram Int 39:1613–1621

Ul Haq E, Kunjalukkal Padmanabhan S, Licciulli A (2014) Synthesis and characteristics of fly ash and bottom ash based geopolymers—a comparative study. Ceram Int 40:2965–2971

Van Jaarsveld J, Van Deventer J (1999) Effect of the alkali metal activator on the properties of fly ash-based geopolymers. Ind Eng Chem Res 38:3932–3941

Van Jaarsveld J, Van Deventer J, Lukey G (2002) The effect of composition and temperature on the properties of fly ash- and kaolinite-based geopolymers. Chem Eng J 89:63–73

Velpari V, Ramachandran BE, Bhaskaran TA, Pai BC, Balasubramanian N (1980) Alkali resistance of fibres in cement. J Mater Sci 15:1579–1584

Venkateshwaran N, Elayaperumal A, Alavudeen A, Thiruchitrambalam M (2011) Mechanical and water absorption behaviour of banana/sisal reinforced hybrid composites. Mater Des 32:4017–4021

Walker R, Pavia S, Mitchell R (2014) Mechanical properties and durability of hemp-lime concretes. Constr Build Mater 61:340–348

Wei J, Meyer C (2015) Degradation mechanisms of natural fiber in the matrix of cement composites. Cem Concr Res 73:1–16

Williams RP, Hart RD, Van Riessen A (2011) Quantification of the extent of reaction of metakaolin-based geopolymers using X-ray diffraction, scanning electron microscopy, and energy-dispersive spectroscopy. J Am Ceram Soc 94:2663–2670

Xu H, Van Deventer J (2000) The geopolymerisation of alumino-silicate minerals. Int J Miner Process 59:247–266

Xu H, Van Deventer JSJ (2002) Geopolymerisation of multiple minerals. Miner Eng 15:1131–1139

Yan L, Chouw N (2015) Effect of water, seawater and alkaline solution ageing on mechanical properties of flax fabric/epoxy composites used for civil engineering applications. Constr Build Mater 99:118–127

Yan S, Sagoe-Crentsil K (2012) Properties of wastepaper sludge in geopolymer mortars for masonry applications. J Environ Manage 112:27–32

Yunsheng Z, Wei S, Zongjin L (2006) Impact behavior and microstructural characteristics of PVA fiber reinforced fly ash-geopolymer boards prepared by extrusion technique. J Mater Sci 41:2787–2794

Yunsheng Z, Wei S, Zongjin L (2010) Composition design and microstructural characterization of calcined kaolin-based geopolymer cement. Appl Clay Sci 47:271–275

Yusuf MO, Megat Johari MA, Ahmad ZA, Maslehuddin M (2014) Influence of curing methods and concentration of NaOH on strength of the synthesized alkaline activated ground slag-ultrafine palm oil fuel ash mortar/concrete. Constr Build Mater 66:541–548

Yusuf M, Megat Johari M, Ahmad Z, Maslehuddin M (2015) Impacts of silica modulus on the early strength of alkaline activated ground slag/ultrafine palm oil fuel ash based concrete. Mater Struct 48:733–741

Zaharaki D, Komnitsas K, Perdikatsis V (2010) Use of analytical techniques for identification of inorganic polymer gel composition. J Mater Sci 45:2715–2724

Zhang ZH, Yao X, Zhu HJ, Hua SD, Chen Y (2009) Preparation and mechanical properties of polypropylene fiber reinforced calcined kaolin-fly ash based geopolymer. J Central South Univ Technol 16:49–52

Zhao FQ, Zhao J, Liu HJ (2009) Autoclaved brick from low-silicon tailings. Constr Build Mater 23:538–541

Zuhua Z, Xiao Y, Huajun Z, Yue C (2009) Role of water in the synthesis of calcined kaolin-based geopolymer. Appl Clay Sci 43:218–223

Chapter 2
Materials and Methodology

Abstract The materials and methodologies used to fabricate and characterise geopolymer composites and nanocomposites are described in this chapter. The characteristics of natural fibers, fabrics and nanofillers used for the fabrication of composites are described. The structure and properties of these materials were investigated using a wide range of techniques which include X-ray diffraction (XRD), Fourier transforms infrared spectroscopy (FTIR), transmission electron microscopy (TEM), scanning electron microscopy (SEM), and thermo-gravimetric analysis (TGA). Methodologies for evaluating flexural strength, flexural modulus, impact strength, fracture toughness, impact toughness, and water absorption are also described.

2.1 Methodology of Synthesis

2.1.1 Cotton Fibre Reinforced Geopolymer Composites

Low calcium fly-ash (ASTM class F), collected from the Collie power station in Western Australia, was used as the source material to prepare the geopolymer composites. The chemical composition of fly ash is shown in Table 2.1. Alkali resistant cotton fibres with an average length of 10 mm, average diameter of 0.2 mm and density of 1.54 g/cm^3 were used to reinforce the geopolymer matrix. The alkaline activator for geopolymerisation was a combination of sodium hydroxide solution and sodium silicate grade D solution. Sodium hydroxide flakes with 98% purity were used to prepare the solution. The chemical composition of sodium silicate used was Na_2O 14.7%, SiO_2 29.4% and water 55.9% by mass.

To prepare the geopolymer composites, the alkaline solution to fly ash ratio of 0.35 was used and the ratio of sodium silicate solution to sodium hydroxide solution was fixed at 2.5. Four samples of geopolymer composites reinforced with 0.3, 0.5, 0.7 and 1 wt% cotton fibre were prepared. Additional water was added to improve the workability and dispersion of cotton fibres in the composite.

An 8-molar concentration of sodium hydroxide solution was prepared and combined with the sodium silicate solution one day before mixing. The fibres were

I.-M. Low et al., *Cotton and Flax Fibre-Reinforced Geopolymer Composites*, Composites Science and Technology,
https://doi.org/10.1007/978-981-16-2281-6_2

Table 2.1 Chemical composition of fly-ash

SiO_2	Al_2O_3	Fe_2O_3	CaO	MgO	SO_3	Na_2O	K_2O	LOI
50%	28.25%	13.5%	1.78%	0.89%	0.38%	0.32%	0.46%	1.64%

added slowly to the dry fly ash in a Hobart mixer at low speed until the mix become homogeneous, at which time the alkaline solution was added. This was mixed for ten minutes on low speed and another ten minutes on high speed. The walls of the mixing container were scraped down to ensure consistency of mix. This procedure was followed for all four test specimens. Each mix was cast in 25 rectangular silicon moulds of 80 mm × 20 mm × 10 mm and placed on a vibration table for five minutes. The specimens were covered with a plastic film and cured at 105 °C for three hours, then rested for 24 h before de-moulding. They were then dried under ambient conditions for 28 days.

2.1.2 Cotton Fabric Reinforced Geopolymer Composites

Cotton fabric (CF) of 30 cm × 7.5 cm was used as a reinforcing material for the fabrication of geopolymer composites. The chemical composition, and the physical properties, of cotton fabric are shown in Tables 2.2 and 2.3 respectively. Low calcium fly-ash (ASTM class F), collected from the Collie power station in Western Australia, was used as the source material of the geopolymer matrix. The chemical compositions of fly-ash are shown in Table 2.1. The alkaline activator for geopolymerisation was a combination of sodium hydroxide solution and sodium silicate grade D solution.

Table 2.2 Chemical analysis of cotton

	Cellulose (%)	Water (%)	Hemicelluloses and pectin (%)	Proteins (%)	Waxes and fats (%)
Cotton fibre	80–90	6–8	4–6	0–1.5	0.5–1

Table 2.3 Properties and structure of cotton fabric

Fabric thickness (mm)	0.41
Fabric geometry	Woven (plain weave)
Yarn nature	Bundle
Filament size (mm)	0.0413
Number of filaments in a bundle	24
Bundle diameter (mm)	0.23
Opening size (mm)	0.5
Fabric density (g/cm^3)	1.6
Tensile strength (MPa)	287–597

Sodium hydroxide flakes of 98% purity were used to prepare the sodium hydroxide solution. The chemical composition of sodium silicate solution was 14.7% Na_2O, 29.4% SiO_2 and 55.9% water by weight.

To prepare the CF-reinforced geopolymer composites the fabric was initially pre-dried for 60 min at 70 °C in an oven. A thin layer of geopolymer matrix was spread into the wooden mould and the first layer of CF was laid upon it and fully impregnated (wet out) with geopolymer paste with a roller before placing the next layer. This process was repeated for the desired number of cotton fibre layers. In each specimen, the final layer was a geopolymer matrix. Pure samples of geopolymer were prepared as controls by slowly adding dry fly ash to the alkaline solution in a Hobart mixer until the mixture became homogeneous. This was then poured into a wooden mould. The alkaline solution to fly-ash ratio was kept at 0.35, and the ratio of sodium silicate solution to sodium hydroxide solution (8 M concentration) was fixed at 2.5.

After casting, each sample was pressed with a 20 kg load for five hours, after which the specimens were covered with plastic film cured at 80 °C in an oven for 24 h and then allowed them to cool down to laboratory conditions before being removed from the mould.

2.1.3 Portland Cement Modified Cotton Fabric Reinforced Geopolymer Composites

Cotton fabric (CF) of 30 cm × 7.5 cm was used as a reinforcing material for the fabrication of geopolymer composites. Low calcium fly-ash (ASTM class F), collected from the Collie power station in Western Australia, was used as the source material of geopolymer matrix, which in this study consisted of fly ash and Ordinary Portland cement. The chemical compositions of fly-ash and Ordinary Portland cement are shown in Table 2.4. The alkaline activator for geopolymerisation was a combination of sodium hydroxide solution and sodium silicate grade D solution. Sodium hydroxide flakes with 98% purity were used to prepare the sodium hydroxide solution. The chemical composition of sodium silicate solution was 14.7% Na_2O, 29.4% SiO_2 and 55.9% water by weight.

Wooden moulds with open tops were prepared and greased to avoid the samples sticking during demoulding. An 8-molar concentration of sodium hydroxide solution was used and added to the sodium silicate solution one day prior to mixing.

The geopolymer matrices were prepared by mixing fly ash and alkaline solutions with three different contents of OPC (i.e., 5, 8 and 10 wt%), using a Hobart mixer.

Table 2.4 Chemical composition of fly-ash (FA) and Ordinary Portland cement (OPC)

Material	SiO_2	Al_2O_3	Fe_2O_3	CaO	MgO (wt%)	SO_3	Na_2O	K_2O	LOI
FA	50	28.25	13.5	1.78	0.89	0.38	0.32	0.46	1.64
OPC	21.10	5.24	3.10	64.39	1.10	2.52	0.23	0.57	1.22

Table 2.5 Compositions of synthesised geopolymer matrix and CF-reinforced geopolymer composites

OPC/geopolymer	OPC content (wt%)	CF-reinforced geopolymer composite	OPC content (wt%)
Pure geopolymer (GP)	0	GP/CF	0
GP/OPC5	5	GP/CF/OPC5	5
GP/OPC8	8	GP/CF/OPC8	8
GP/OPC10	10	GP/CF/OPC10	10

The final mixture was poured into wooden moulds and covered with plastic film and allowed to cure for 24 h at ambient temperature before de-moulding. Subsequently, the mixtures were left for 28 days at room temperature. Pure geopolymer (GP) samples without OPC addition was also made, as a control.

In the preparation of CF reinforced geopolymer composites the fabric was initially pre-dried for 60 min at 70 °C in an oven. A thin layer of geopolymer matrix was spread into the wooden mould and the first layer of CF was carefully laid upon it and fully impregnated (wet out) with geopolymer paste with a roller before placing the next layer. This process was repeated for the desired number of cotton fibre layers. In each specimen, the final layer was geopolymer matrix. The alkaline solution to fly-ash ratio was kept at 0.35, while the ratio of sodium silicate solution to sodium hydroxide solution was fixed at 2.5. After casting, each sample was pressed with a 20 kg load for five hours, after which the specimens were covered with plastic film and allowed to cure for 24 h at ambient condition before de-moulding. All specimens were then kept under ambient condition for 28 days.

The same process was used to prepare the composites without the addition of OPC. The amount of CF in the final products was approximately 17.2 wt% (three layers). All the samples made are summarised in Table 2.5.

2.1.4 *Flax-Fabric Reinforced Geopolymer Composites*

Low calcium fly ash (ASTM class F), collected from the Eraring power station in NSW, was used as the source material for the geopolymer matrix. The chemical composition of fly ash is shown in Table 2.1. The alkaline activator for geopolymerisation was a combination of sodium hydroxide and sodium silicate grade D solution. Sodium hydroxide flakes of 98% purity were used to prepare the solution. The chemical composition of sodium silicate used was 14.7% Na_2O, 29.4% SiO_2 and 55.9% water by mass.

Flax fabric (FF) and organo-nanoclay (Cloisite 30B) were used for the reinforcement of geopolymer nanocomposites. The fabric of 30×30 cm^2, supplied by Pure Linen Australia, is made up of yarns with a density of 1.5 g/cm^3; the space between

Table 2.6 Structure and physical properties of the flax fabric

Fabric thickness (mm)	0.6
Fabric geometry	Woven (plain weave)
Yarn nature	Bundle
Bundle diameter (mm)	0.6
Filament size (mm)	0.01–0.02
Opening size (mm)	2–4
Fabric density (g/cm^3)	1.5
Modulus of elasticity (GPa)	39.5
Tensile strength (MPa)	660

Table 2.7 Physical properties of the organo-nanoclay platelets (Cloisite 30B)

Colour	Off white
Density (g/cm^3)	1.98
d-spacing (001) (nm)	1.85
Aspect ratio	200–1000
Surface area (m^2/g)	750
Typical dry particle sizes	90% volume < 13 μm
	50% volume < 6 μm
	10% volume < 2 μm

the yarns is between 2 and 4 mm, necessary to allow the geopolymer matrix to penetrate. The average diameter of the fibre yarns was 0.60 mm, and the fibres diameter was about 20 μm. The physical properties of the flax fibres are presented in Table 2.6. The nanoclay platelets used in this study was based on natural montmorillonite clay ($(Na,Ca)_{0.33}(Al,Mg)_2(Si_4O_{10})(OH)_2.nH_2O$) which was supplied by Southern Clay Products, USA. The description and physical properties of Cloisite 30B are shown in Table 2.7 (Hakamy et al. 2015).

To prepare the geopolymer matrix, an alkaline solution to fly ash ratio of 0.75 was used and the ratio of sodium silicate solution to sodium hydroxide solution was fixed at 2.5. The concentration of sodium hydroxide solution was 8 M and was prepared and combined with the sodium silicate solution one day before mixing.

The nanoclay was added first to the fly ash at the dosages of 0, 1.0, 2.0 and 3.0% by weight. The fly ash and nanoclay were dry mixed for 5 min in a covered mixer at a low speed and then mixed for another 10 min at high speed until homogeneity was achieved. The alkaline solution was then added slowly to the fly ash/nanoclay in the mixer at a low speed until the mix became homogeneous, then further mixed for another 10 min on high speed. The resultant mixture was then poured into wooden moulds and placed on a vibration table for two minutes.

Similar mixtures were prepared to produce the nanocomposites reinforced with FF. Four samples of geopolymer pastes reinforced with 4.1 wt% FF were prepared by spreading a thin layer of geopolymer paste in a well-greased wooden mould and carefully placing the first layer of FF on it. The fabric was fully saturated with paste by

Table 2.8 Formulation of samples

Sample name	Fly-ash (g)	NaOH solution (g)	Na_2SiO_3 solution (g)	Nanoclay (g)	FF content (wt%)
GP	1000	214.5	535.5	0	0
GPNC-1	1000	214.5	535.5	10	0
GPNC-2	1000	214.5	535.5	20	0
GPNC-3	1000	214.5	535.5	30	0
GPFNC-0	1000	214.5	535.5	0	4.1
GPFNC-1	1000	214.5	535.5	10	4.1
GPFNC-2	1000	214.5	535.5	20	4.1
GPFNC-3	1000	214.5	535.5	30	4.1

a roller, and the process repeated for ten layers; each specimen contained a different weight percentage of nanoclay. The samples then were left under heavy weight for 1 h to reduce entrapped air inside the samples. All samples were covered with plastic film and cured at 80 °C for 24 h in an oven before demoulding. They were then dried under ambient conditions for 28 days. The pure geopolymer, and nanocomposites containing 1.0%, 2.0% and 3.0% nanoclay were labelled as GP, GPNC-1, GPNC-2 and GPNC-3, respectively. Also, the composites reinforced with a combination of FF and the same weight percentages of nanoclay were denoted as GPFNC-0, GPFNC-1, GPFNC-2 and GPFNC-3, respectively (see Table 2.8).

2.1.5 Flax Fabric Reinforced Geopolymer Nanocomposites

Figure 2.1 provides a summary of the process followed in preparing all samples. To prepare the pure geopolymer and geopolymer nanocomposites, nano silica was

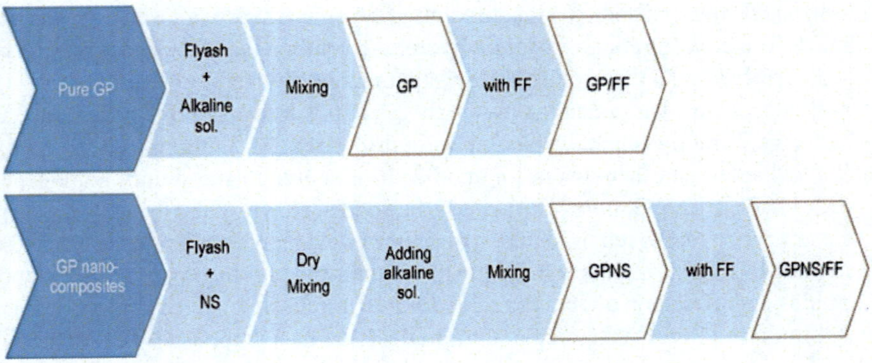

Fig. 2.1 Diagram presenting the procedures of producing the samples

added to the fly ash at 0.0, 0.5, 1.0, 2.0 and 3.0% by weight. They were placed in a covered mixer to dry-mix at a low speed for 5 min. The dry-mixing method was chosen as it ensures better dispersion of the nanoparticles with fly ash powder. They were then left for an additional 10 min to mix at high speed until the homogeneity of dry mix was obtained. The next step was addition of the alkaline solution to the dry mix and mixed initially at low speed awaiting the mixture to achieve homogeneity and then for an additional ten minutes at high speed. The mixture that resulted was poured in to wooden moulds which were then vibrated for 2 min on a vibrating table before covering them with a plastic sheet. The moulds were then placed in oven for 24 h at 80 °C, where geopolymer reaction happened. The geopolymer paste without nano silica particles was considered as the control sample.

Fabrication of the flax fabric (FF) reinforced composites and the composites containing nano silica (here after called as nanocomposites) requires preparation of similar mixes. Four samples of geopolymers were reinforced using ten layers of flax fabric. The process required the paste being spread into a greased wooden mould followed by the first layer of flax fabric. Then a roller was used to completely saturate the fabric with the paste. The process was done repeatedly for ten layers. To ensure the removal of entrapped air inside the samples, a 20 kg weight was placed on top of the samples. A plastic sheet was used to cover each of samples, and then they were cured (at 80 °C) for 24 h in an oven before demoulding. The resulting samples were then dried for 28 days under ambient conditions. The geopolymer matrix that was reinforced using flax fabric without nano silica is considered as control sample.

2.2 Techniques of Characterisation

2.2.1 X-Ray Diffraction and X-Ray Fluorescence

Identical pieces have been selected as well as cut from the samples prepared. They were then crushed before grounding to fine powder. A D8 Advance Diffractometer was used to measure the samples of powder through use of copper radiation as well as a position sensitive detector (the LynxEye). At 0.5°/min (in scanning rate) scanning of the diffractometer was done from 7.5° to 60°. To get XRD patterns, Cu k_{α} lines (k = 1.5406 Å) were utilized. Additionally, *MAUD* V2.44 software was used to perform QXDA (Quantitative X-ray Diffraction Analysis) with Rietveld refinement. The compound selected to represent an internal standard was a fluorite (CaF_2) (Rickard et al. 2011). To prepare the QXDA samples, geopolymer paste (3.0 g in dry weight) was mixed with fluorite (0.33 g in weight). Parameters that were defined based on Rietveld were then used to calculate the weight percentage of the individual crystalline phases (W_{Cr}) through Eq. 2.1 (Chen-Tan et al. 2009) shown below:

$$W_{Cr} = \left[W_{std} \frac{S_{Cr}(ZMV)_{Cr}}{S_{std}(ZMV)_{std}} \right] \times \left[\frac{1}{1 - W_{std}} \right] \quad (2.1)$$

W_{std} represents the standard weight per cent, M represents the mass of unit cells while V is their volume. Z represents the number of units of the formula for every unit cell. S_{Cr} represents the scale factor for the crystalline phases, and S_{std} represents the scale factor for the standard. Calculation of the amorphous weight content W_{Am} was then done based on the equation below (Chen-Tan et al. 2009):

$$W_{Am} = 1 - \sum_{i=1}^{n} W_n \quad (2.2)$$

where n denotes the number of crystalline phases that were refined.

The chemical compositions of Eraring fly ash were analysed using X-ray fluorescence (XRF). FTIR scan was conducted using a FTIR spectrometer (the Perkin Elmer Spectrum 100) on the range 4000–500 cm^{-1} range and at room temperature. The samples were examined to determine fracture surfaces and microstructures by using SEM (the Zeiss EVO-40 from Carl Zeiss, Germany). The samples of geopolymer were put in a vacuum desiccator for two days seeking to complete out-gassing before mounting the samples on aluminium stubs followed by coating with a thin platinum layer.

2.2.2 *Thermogravimetric Analysis (TGA)*

Thermogravimetric analysis (TGA) was carried out for cotton fibre, flax fibre, unreinforced geopolymer and reinforced geopolymer composites at a heating rate of 10 °C/min under atmospheric condition. The temperature range scanned between 50 and 1000 °C. The weight of all specimens was maintained around 15 mg.

2.2.3 *Scanning Electron Microscopy (SEM)*

A Zeiss Evo 40XVP scanning electron microscope was used to examine the microstructures of fly-ash and geopolymer composites. The specimens were mounted on aluminium stubs using carbon tape, and then coated with a thin layer of platinum to prevent charging before the observation.

2.2.4 *Fourier Transform Infrared (FTIR) Spectra*

The Fourier transform infrared spectroscopy (FTIR) was performed on Perkin Elmer Spectrum 100 FTIR spectrometer in the transmission mode at room temperature. FT-IR spectra were recorded in the range (4000–500 cm^{-1}) at a resolution of 2 cm^{-1} with 10 scans. Background spectra were taken in the empty chamber before measurements to eliminate the influence of water moisture and CO_2 in air.

2.3 Physical and Mechanical Properties

2.3.1 *Density and Porosity*

Both density and porosity tests were performed to determine the quality of each geopolymer composite sample. The values of bulk density (D) and apparent porosity (P_s) were determined in accordance with the ASTM Standard (C-20) and calculated using the following equations (ASTM C20 2010):

$$D = \frac{W_d}{W_a - W_w} \tag{2.3a}$$

$$P_s = \frac{W_a - W_d}{W_a - W_w} \times 100 \tag{2.3b}$$

where W_d = weight of the dried sample, W_w = weight of the sample saturated with and suspended in water, and W_a = weight of the sample saturated in air.

2.3.2 *Moisture Absorption*

The composite specimens used for moisture absorption test were immersed in a water bath at room temperature for longer period to reach equilibrium. At regular intervals, the specimens were taken out from the water and wiped with filter paper to remove surface water and weighed with digital scale (AA-200, Denver Instrument Company, USA). The samples were re-immersed in water to permit the continuation of sorption until saturation limit was reached after 133 days. The weighing was done within 30 s, to avoid the error due to evaporation. The percentage of the water content (M_t) was determined using the following equation (Kim et al. 2005):

$$M_t\,(\%) = \left(\frac{W_t - W_o}{W_o}\right) \times 100 \tag{2.4a}$$

where W_t is the weight of the sample at time t and W_o is the initial weight of the sample.

The water absorption behaviour in the samples can be studied as Fickian behaviour. Therefore, the following formula has been used (Mohan and Kanny 2011):

$$\frac{M_t}{M_\infty} = 4\left(\frac{Dt}{\pi h^2}\right)^{1/2} \quad (2.4b)$$

where M_t is the water content at time t, M_∞ is the equilibrium water content, D is the diffusion coefficient and h is the sample thickness.

2.3.3 *Flexural Strength and Modulus*

Rectangular bars with a length of 40 mm were cut from the fully cured samples and subjected to three-point bend tests to evaluate their flexural strength and modulus. A LLOYD Material Testing Machine (50 kN capacity) with a displacement rate of 1.0 mm/min was employed to perform the tests. In total, five specimens of each composition were tested. The flexural strength (σ_F) was determined using the following equation:

$$\sigma_F = \frac{3}{2}\frac{P_m S}{B W^2} \quad (2.5)$$

where P_m is the maximum load at crack extension, S is the span of the sample, B is the specimen width and W is the specimen thickness or depth. The flexural modulus was computed using the initial slope of the load–displacement curve, $\Delta P/\Delta X$, using the following formula:

$$E_F = \frac{S^3}{4WD^3}\left(\frac{\Delta P}{\Delta X}\right) \quad (2.6)$$

2.3.4 *Impact Strength*

A Zwick Charpy impact tester with a 1.0 J pendulum hammer was employed to determine the impact strength. For each composition, five bars of 40 mm length were tested. The impact strength (σ_i) was calculated using the following equation:

$$\sigma_i = E/A \quad (2.7)$$

where E is the impact energy required to break a sample with a ligament of area A.

2.3.5 Rockwell Hardness

The hardness of geopolymer composites was measured using an Avery Rockwell hardness tester at hardness scale H. Before measurement, the surfaces of test samples were polished using a Struers Pedimat polisher, finishing with 10 μm grade diamond paste.

2.3.6 Fracture Toughness

Rectangular bars of 80 mm in length with a cross-sectional dimension of 20 × 20 mm were used in fracture toughness measurements. Subsequently, a crack with a length to thickness (depth) (a/W) ratio of 0.4 was introduced in each specimen by means of a 0.4 mm diamond blade. The fracture toughness (K_{IC}), was calculated using the equation proposed by Low et al. (2007):

$$K_{IC} = \frac{p_m S}{B W^{3/2}} f\left(\frac{a}{W}\right) \tag{2.8a}$$

where P_m is the maximum load at crack extension, S is the span of the sample, B is the specimen width, W is the specimen thickness (depth), a is the crack length and f(a/W) is the polynomial geometrical correction factor given by the equation below (Low et al. 2007):

$$f\left(\frac{a}{W}\right) = \frac{3(a/W)^{1/2}[1.99 - (a/W)(1 - a/W) \times (2.15 - 3.93a/W + 2.7a^2/W^2)]}{2(1 + 2a/W)(1 - a/W)^{3/2}} \tag{2.8b}$$

2.3.7 Flexural Toughness and Toughness Indices

The flexural toughness of FF-reinforced nanocomposites was characterised and evaluated by the toughness indices I_5, I_{10} and $I_{failure}$ as defined by ASTM C1018 (1998), Standard Test Method for Flexural Toughness and First-Crack Strength of Fiber-Reinforced Concrete (Using Beam with Third-Point Loading). According to the standard, I_5 is defined as the ratio obtained by dividing the area up to a deflection of three times the first-crack deflection by the area up to first crack deflection, while I_{10} is the ratio of the area up to a deflection of 5.5 times the first-crack deflection to the area up to the first crack (see Fig. 2.2). Thus, it could be written as:

$$I_5 = \frac{(Area)_{OACDO}}{(Area)_{OABO}} \tag{2.9}$$

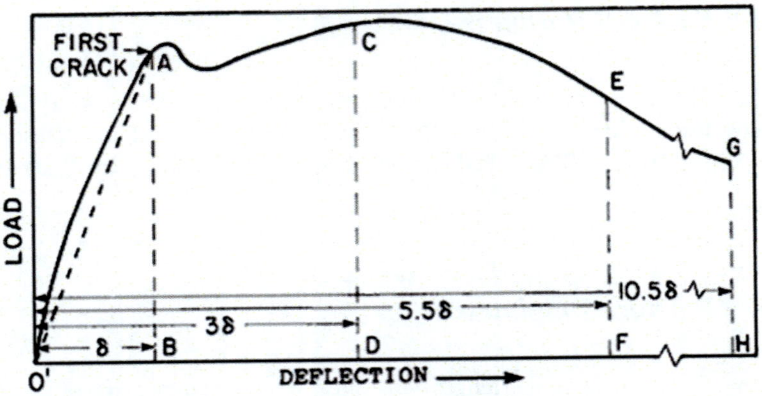

Fig. 2.2 Definition of ASTM toughness indices (ASTM C1018 1998)

$$I_{10} = \frac{(Area)_{OAEFO}}{(Area)_{OABO}} \quad (2.10)$$

Similarly, $I_{failure}$ can be evaluated for the failure deflection of 11.4 mm in this study as most of the specimens lost their load carrying capacity significantly at that deflection.

References

ASTM C1018 (1998) Standard test method for flexural toughness and first-crack strength of fiber-reinforced concrete (using beam with third-point loading)

ASTM C20 (2010) Standard test methods for apparent porosity, water absorption, apparent specific gravity, and bulk density of burned refractory brick and shapes by boiling water

Chen-Tan NW, Van Riessen A, Ly CV, Southam DC (2009) Determining the reactivity of a fly ash for production of geopolymer. J Am Ceram Soc 92:881–887

Hakamy A, Shaikh FUA, Low IM (2015) Characteristics of nanoclay and calcined nanoclay-cement nanocomposites. Compos B Eng 78:174–184

Kim JK, Hu C, Woo RSC, Sham ML (2005) Moisture barrier characteristics of organoclay-epoxy nanocomposites. Compos Sci Technol 65:805–813

Low IM, Mcgrath M, Lawrence D, Schmidt P, Lane J, Latella BA, Sim KS (2007) Mechanical and fracture properties of cellulose-fibre-reinforced epoxy laminates. Compos A Appl Sci Manuf 38:963–974

Mohan TP, Kanny K (2011) Water barrier properties of nanoclay filled sisal fibre reinforced epoxy composites. Compos A Appl Sci Manuf 42:385–393

Rickard WDA, Williams R, Temuujin J, Van Riessen A (2011) Assessing the suitability of three Australian fly ashes as an aluminosilicate source for geopolymers in high temperature applications. Mater Sci Eng A 528:3390–3397

Chapter 3
Physical Properties

Abstract This chapter describes the physical properties of geopolymer reinforced with natural fibres and/or nanofillers. Results show that the appropriate addition of natural fibres and/or nanofillers can modify the microstructures and thus the attending physical properties of geopolymer composites and nanocomposites.

3.1 Cotton Fibre-Reinforced Geopolymer Composites

3.1.1 Synchrotron Radiation Diffraction

The synchrotron radiation diffraction (SRD) patterns of commercial fly ash, cotton fibres and of prepared geopolymer reinforced with 0.3, 0.5, 0.7 and 1.0 wt% of cotton fibres are shown in Fig. 3.1. The diffraction pattern of cotton fibres shows typical characteristic peaks, indicating the presence of cellulose. Fly ash displays peaks caused by the presence of quartz and mullite as well as other crystalline phases. In addition, a broad peak, can be discerned in the region around $2\theta = 30°$, arising from the amorphous phase present. This amorphous phase is crucial for geopolymerisation reactions (Rattanasak and Chindaprasirt 2009) which lead to the formation of a geopolymer (Rickard et al. 2011).

Comparing the SRD spectra of the original fly ash with those of the hardened geopolymeric composites, Fig. 3.1 indicates that the crystalline phases (quartz, mullite, etc.) originally existed in the fly ash have apparently not been altered by the activation reactions; hence they do not participate in the geopolymerisation reaction. The diffraction patterns of geopolymer composites reinforced with 0, 0.3, 0.5, 0.7 and 1 wt% cotton fibres all showed the sharp peaks of the crystalline phases from fly ash, thus confirming that these phases are neither reactive nor involved in geopolymerisation but are simply present as inactive fillers in the geopolymer network.

I.-M. Low et al., *Cotton and Flax Fibre-Reinforced Geopolymer Composites*, Composites Science and Technology,
https://doi.org/10.1007/978-981-16-2281-6_3

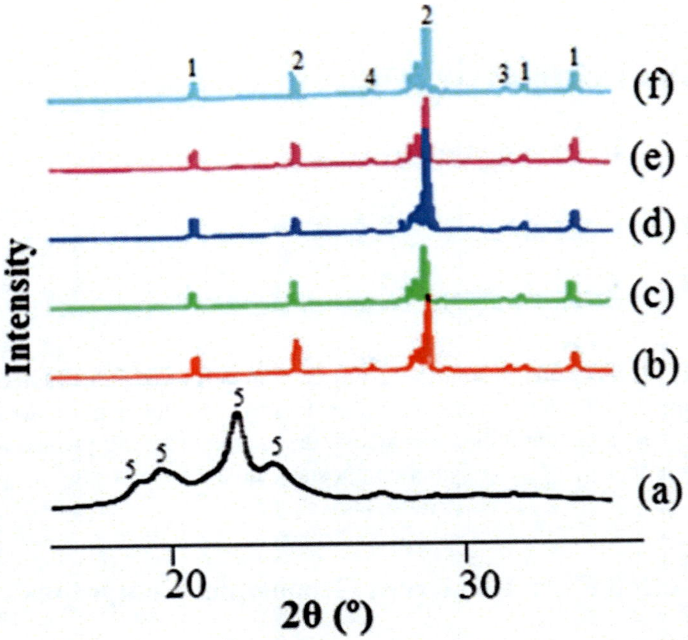

Fig. 3.1 Synchrotron radiation diffraction patterns of **a** cotton fibres (CF), **b** fly-ash, and geopolymer composite with **c** 0.3 wt% CF, **d** 0.5 wt% CF, **e** 0.7 wt% CF, and **f** 1.0 wt% CF. Legend: 1 = mullite, 2 = quartz, 3 = maghemite, 4 = hematite, 5 = cellulose

3.1.2 Density and Porosity

The density and porosity values of the geopolymer composites after 28 days of curing at ambient temperature are presented in Figs. 3.2 and 3.3, respectively. Figure 3.2

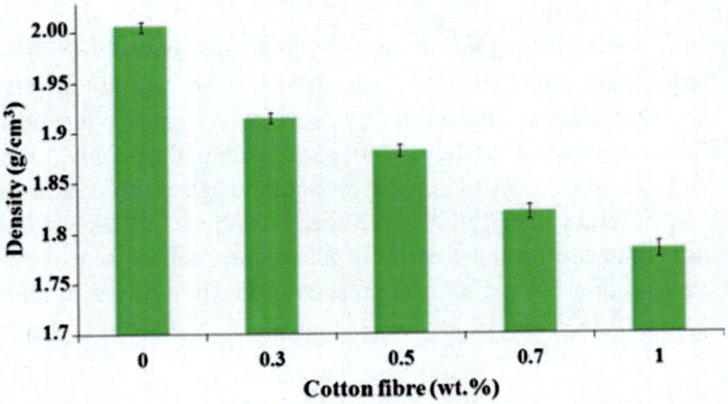

Fig. 3.2 Density of geopolymer composites as a function of fibre content

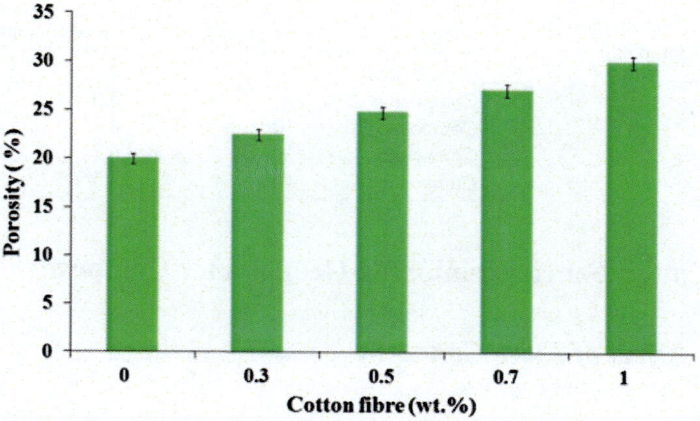

Fig. 3.3 Porosity of geopolymer composites as a function of fibre content

shows that density decreases as the weight percent of cotton fibre increases. The geopolymer composite reinforced with 1.0 wt% of cotton fibre has the lowest density of 1.8 g/cm^3 whereas the control sample displays the highest value of ~2.0 g/cm^3. These results agree with those obtained by other investigators (Aggarwal 1995; Abdullah et al. 2011). For instance, the study on bagasse fibre-reinforced cement composite reported that the density values decreased with increase of fibre content (Aggarwal 1995). Similarly, in another study by Abdullah et al. (2011) on coconut fibre reinforced cement, they reported that density values of cement composites decreased with increasing fibre content.

The value of porosity increases with increases in the weight percent of cotton fibres as shown in Fig. 3.3. The lowest value of porosity (20%) is found in the control sample that contained no cotton fibres whereas the composite containing the highest amount of cotton fibre has the highest porosity of 30%.

The effect of the initial water content on density and porosity has perhaps the most important implications in this study. To reduce the viscosity of the geopolymer composites with 0.7 and 1.0 wt% of cotton fibres, a high water/fly ash ratio was required, and this caused an increase of porosity in the resulting composites. The addition of extra water results in larger amounts of "free" water that is trapped in inter-granular space or large pores after geopolymerisation and evaporates during curing and extended ageing, leaves large quantities of inter-granular pores in the microstructure (Criado et al. 2012; Kouamo et al. 2012).

The increase in porosity with increasing cotton fibre content may also be explained by the fact of water absorbed by the fibres. It is possible that fibres tend to clump together during mixing, entrapping water-filled spaces that subsequently turn into voids. Thus, increased fibre content may enhance the potential for fibre clumping which is undesirable for achieving a uniform microstructure (Neithalath et al. 2004).

Table 3.1 Formulations of fabricated samples

Sample	Fabric layers	Fibre content (wt%)
Composite 0	0	0
Composites 1	5	3.6
Composites 2	10	4.5
Composites 3	20	6.2
Composites 4	40	8.3

3.2 Cotton Fabric-Reinforced Geopolymer Composites

3.2.1 *Effect of Fibre Content*

The compositions of the fabricated samples are shown in Table 3.1.

Synchrotron Radiation Diffraction

The synchrotron radiation diffraction (SRD) patterns of class F fly ash, cotton fibres, and geopolymer composites containing 0, 3.6, 4.5, 6.2 and 8.3 wt% of cotton fibres, are shown in Fig. 3.4. The crystalline phases present were indexed using Powder Diffraction Files (PDFs) from the Inorganic Crystal Structure Database (ICSD). The diffraction pattern of cotton fibres shows typical characteristic peaks, indicating the presence of cellulose. The SRD pattern of fly ash shows that the major crystalline phases are quartz, mullite and hematite. These crystalline phases are the main component of fly ash. When comparing the SRD pattern of the original fly ash with those of the hardened geopolymeric materials shown in Fig. 3.4, it is seen that the crystalline phases of quartz, mullite and hematite remain unchanged and have not been visibly altered by the activation reaction. This finding confirms that the crystalline phases are not reactive or involved in geopolymerisation, but simply present as inactive fillers in geopolymer network (Fernández-Jiménez and Palomo 2003, 2005; Rattanasak and Chindaprasirt 2009; Rickard et al. 2011). In this case, however, ***amorphous aluminosilicate*** phases are more reactive and dissolvable in alkaline solution during the formation of a geopolymer (Chen-Tan et al. 2009; Xu and Van Deventer 2000, 2002).

3.2.2 *Effect of Nanoclay*

Density, Porosity and Water Absorption

Table 3.2 shows the compositions of the fabricated samples. The results of porosity and water absorption of geopolymer paste and geopolymer nanocomposites are shown in Table 3.3. All geopolymer nanocomposites showed higher densities and lower porosities than the control paste. The addition of nano-clay has increased the density and reduced the porosity and the water absorption of geopolymer nanocomposites when compared to control geopolymer paste. The optimum addition was

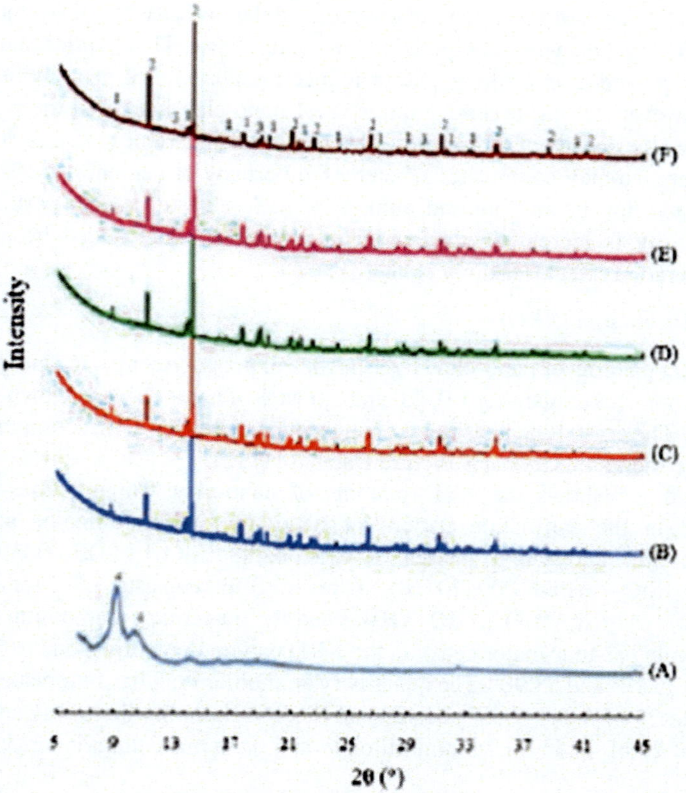

Fig. 3.4 Synchrotron radiation diffraction patterns of **a** cotton fibres (CF), **b** fly-ash, and geopolymer composite with **c** 3.6 wt% CF, **d** 4.5 wt% CF, **e** 6.2 wt% CF and **f** 8.3 wt% CF. Legend: 1 = mullite (PDF 15–0776), 2 = quartz (PDF 05–0490), 3 = hematite (PDF 13–0534), 4 = cellulose (PDF 00–060–1502)

Table 3.2 Formulation of samples

Sample	Fly-ash (g)	NaOH solution (g)	Na_2SiO_3 solution (g)	Nano-clay (g)
GP	1000	214.5	535.5	0
GPNC-1	1000	214.5	535.5	10
GPNC-2	1000	214.5	535.5	20
GPNC-3	1000	214.5	535.5	30

Table 3.3 Porosity and water absorption for pure geopolymer and geopolymer nano-composites

Sample	Density (g/cm^3)	Porosity (%)	Water absorption (%)
GP	1.84 ± 0.02	22.2 ± 0.4	12.1 ± 0.2
GPNC-1	1.92 ± 0.02	21.3 ± 0.3	11.1 ± 0.1
GPNC-2	2.05 ± 0.02	20.6 ± 0.3	10.0 ± 0.2
GPNC-3	1.98 ± 0.03	21.0 ± 0.2	10.6 ± 0.2

found as 2.0 wt% of nano-clay, which reduced the porosity by 7.1%, and the water absorption by 17% when compared to the control paste. This implies that nano-clay particles played a pore-filling role to reduce the porosity of geopolymer composites. However, adding excessive amounts of nano-clay increased the porosity and decreased the density of all samples. This result is comparable to that of cement reinforced organo-clay composites whereby the porosity of cement paste is decreased due to addition of an optimum amount of nano-clay to cement paste. However, the porosity is increased when more nanoparticles were added because of the agglomeration effect (Hakamy et al. 2013).

X-Ray Diffraction (XRD)

The XRD patterns of nano-clay, fly ash, control geopolymer paste, and geopolymer nanocomposites containing 1.0, 2.0 and 3.0 wt% of nano-clay are shown in Figs. 3.5 and 3.6. The crystalline phases were indexed using Powder Diffraction Files (PDFs) from the Inorganic Crystal Structure Database (ISCD).

Figure 3.5 shows the XRD patterns of nano-clay. Three phases have been indexed in the diffraction pattern of nano-clay with the major phase being Cloisite 30B and minor phases of Cristobalite [SiO_2] (PDF-000391425) and Quartz [SiO_2] (PDF-000470718). Cloisite 30B consists of Montmorillonite [$(Ca,Na)_{0.3}Al_2(Si,Al)_4O_{10}(OH)_2 \cdot xH_2O$] and the quaternary ammonium salt. Montmorillonite has four major peaks in the XRD pattern that correspond to 2θ of 4.84°, 19.74°, 35.12° and 53.98°. The quaternary ammonium salt has four peaks that correspond to 2θ of 4.84°, 9.55°, 24.42° and 29.49°. Note that there was an overlap of peaks at 2θ of 4.84° for Montmorillonite and quaternary ammonium salt. Both of

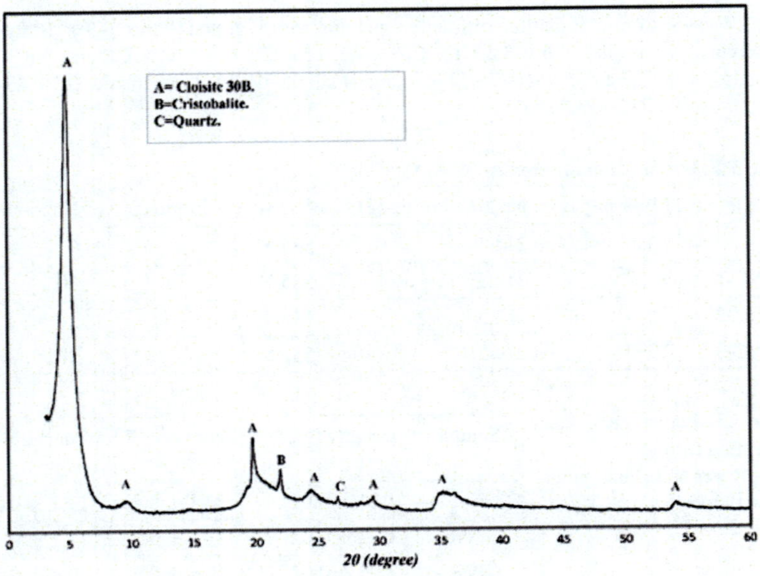

Fig. 3.5 X-ray diffraction pattern of nano-clay (Cloisite 30B)

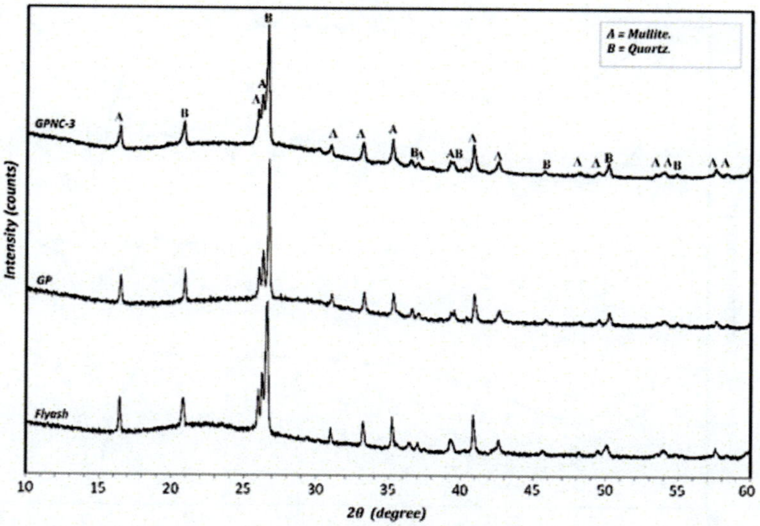

Fig. 3.6 X-ray diffraction patterns of fly-ash, GP and GPNC-3

Cristobalite and Quartz has a peak that corresponds to 2θ of 21.99° and 26.61° respectively. The broad hump in the diffraction pattern of Cloisite 30B indicates the presence of amorphous content in the nanoclay.

Figure 3.6 shows two important phases: quartz [SiO_2] (PDF-010872096) and mullite [$Al_{4.56}$ $Si_{1.44}$ $O_{9.72}$] (PDF-010791458). These crystalline phases are mainly the fly-ash phases, and they are not reactive in the geopolymeric reaction, but they are existing as unreactive and filler particles in the geopolymer paste (Fernández-Jiménez and Palomo 2005; Alomayri and Low 2013a, b). However, the amorphous aluminosilicate phase generated between $2\theta = 14°$ and 27° is a sign of the activity of geopolymeric reaction, which is the reactive and dissolvable content in alkaline solution throughout the geopolymer formation (Chen-Tan et al. 2009). This amorphous phase affects the mechanical properties of geopolymer matrix significantly: the higher the content of amorphous phase, the higher the strength exhibited by the geopolymer (Bakharev 2006; Rickard et al. 2011).

FTIR Observation

FTIR spectra of both pure geopolymer and geopolymer nanocomposite are shown in Fig. 3.7. The FTIR spectra of all samples shows a strong peak at ~1000 cm^{-1} which is associated with Si–O–Si asymmetric stretching vibrations and is the fingerprint of the geopolymerisation (Phair and Van Deventer 2002). A broad peak in the region around 3340 cm^{-1} is corresponding to the hydroxyl (OH) group of physically free water (higher frequencies), and to chemically bounded water through hydrogen bonds (lower frequencies) (Alamri and Low 2013). The absorbance peak at 1640 cm^{-1} is also attributed to the (OH) bending vibration (Ul Haq et al. 2014). The band at 1440 cm^{-1} is an indicator of the presence of sodium carbonate; this

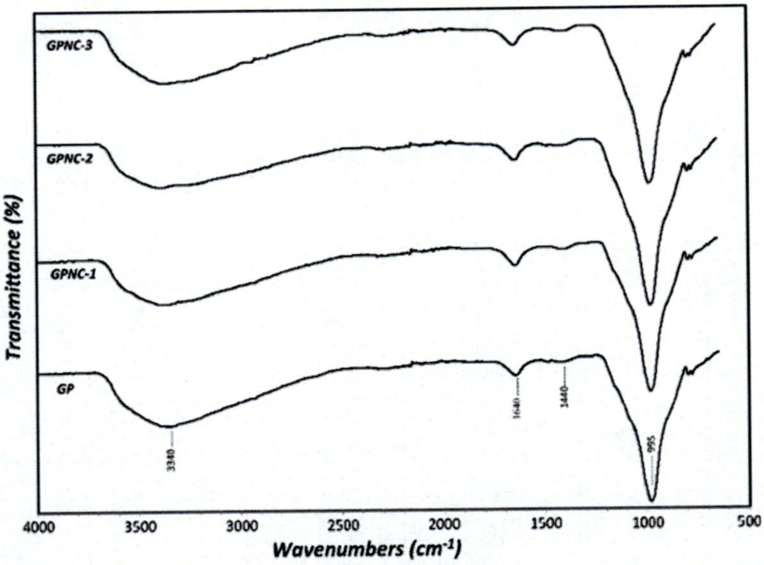

Fig. 3.7 FTIR spectra of all samples

Table 3.4 Peak areas and peak heights ratios of geopolymers at Si–O–Si stretching vibrations from FTIR spectra

Sample	Wave-number of Si–O–Si peak	Ratio of peak heights	Ratio of peak areas
GP	983	1	1
GPNC-1	983	1.02	1.02
GPNC-2	981	1.19	1.07
GPNC-3	980	1.15	1.03

was produced because of the atmospheric carbonation on the surface of the matrix where it reacts with carbon dioxide (Zaharaki et al. 2010). The level of geopolymerization can be specified by measuring the ratios of the height and the area of the Si–O–Si stretching peaks of the nanocomposites to the pure matrix (Ul Haq et al. 2014). Table 3.4 illustrates that all nanocomposites had generally higher contents of geopolymer compared to the control paste; however, the addition of 2.0 wt% of nano-clay had the highest level of geopolymerization among all samples.

3.2.3 *Effect of Ordinary Portland Cement*

The chemical compositions of fly-ash and Ordinary Portland cement are shown in Table 3.5. The alkaline activator for geopolymerisation was a combination of sodium hydroxide solution and sodium silicate grade D solution. Sodium hydroxide flakes

Table 3.5 Chemical composition of fly-ash (FA) and Ordinary Portland cement (OPC)

Material	SiO_2	Al_2O_3	Fe_2O_3	CaO	MgO (wt%)	SO_3	Na_2O	K_2O	LOI
FA	50	28.25	13.5	1.78	0.89	0.38	0.32	0.46	1.64
OPC	21.10	5.24	3.10	64.39	1.10	2.52	0.23	0.57	1.22

with 98% purity were used to prepare the sodium hydroxide solution. The chemical composition of sodium silicate solution was: 14.7% Na_2O, 29.4% SiO_2 and 55.9% water by weight. The compositions of samples made have been tabulated in Table 3.6.

XRD Analysis

Figures 3.8 and 3.9 show the XRD analysis results for the Portland cement and fly ash powders and the geopolymer specimens prepared with 0 and 10 wt% OPC. It can be seen from Fig. 3.8 that the XRD pattern of the OPC powder represents many important phases in this study: portlandite [$Ca(OH)_2$] (PDF 00-044-1481), dicalcium silicate [C_2S] (PDF 00-033-0302), tricalcium silicate [C_3S] (00-049-0442), Ettringite [$Ca_6Al_2(SO_4)_3(OH)_{12}.26H_2O$] (PDF 000411451), Gypsum [$Ca(SO_4)(H_2O)_2$] (PDF 040154421), Quartz [SiO_2] (PDF 000461045) and Calcite [$CaCO_3$] (PDF 000050586). The fly ash powder is mainly consisting of mullite (PDF 015-0776) and quartz (PDF 005-0490) as well as other crystalline phases as shown in Fig. 3.9a. These phases are not involved in the geopolymerisation reaction, but the amorphous phase generated by coal combustion. The amorphous phase is crucial for the geopolymerisation reactions which occur at this stage lead to the formation of a geopolymer

Table 3.6 Compositions of synthesised geopolymer and CF-reinforced geopolymer composites

Geopolymer/OPC	OPC content (wt%)	CF-reinforced geopolymer composite	OPC content (wt%)
Pure geopolymer (GP)	0	GP/CF	0
GP/OPC5	5	GP/CF/OPC5	5
GP/OPC8	8	GP/CF/OPC8	8
GP/OPC10	10	GP/CF/OPC10	10

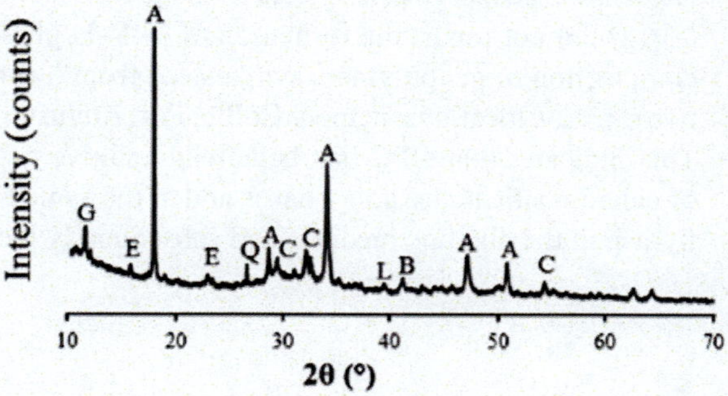

Fig. 3.8 XRD pattern of Ordinary Portland cement powder. Legend: A = $Ca(OH)_2$, B = C_2S, C = C_3S, E = ettringite, G = gypsum, Q = quartz, L = calcite

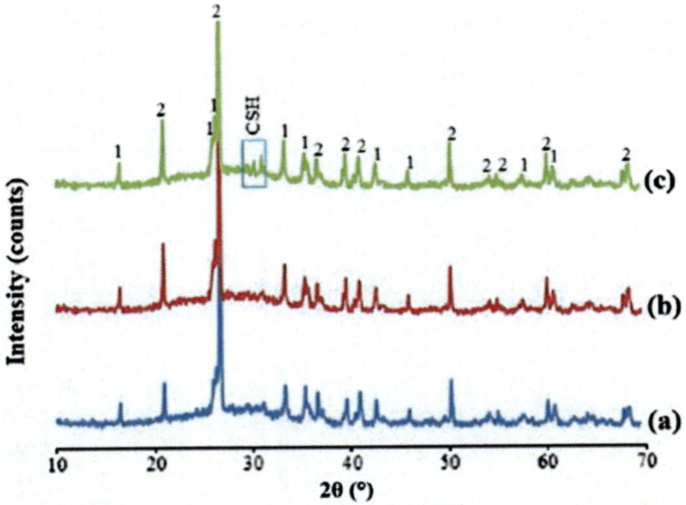

Fig. 3.9 XRD patterns of **a** fly ash and geopolymer specimens prepared with **b** 0 wt% OPC and **c** 10 wt% OPC. Legend: 1 = mullite, 2 = quartz, CSH = calcium silicate hydrate

(Rattanasak and Chindaprasirt 2009; Rickard et al. 2011). Comparing the XRD of the original fly ash with that of the hardened geopolymeric composites (0 wt% OPC), plots (a) and (b) in Fig. 3.9 indicate that the crystalline phases (quartz, mullite, etc.) originally existing in the fly ash have apparently not been altered by the activation reactions; hence, they do not participate in the geopolymerisation reaction.

There is a new phase (calcium silicate hydrate) [C–S–H] (PDF 014-0035) formed because of the hydration reaction in the presence of OPC; this observed C–S–H phase has been reported by other researchers (Yip and van Deventer 2003; Yip et al. 2005). Buchwald et al. (2007) observed the formation of C–S–H gel in the slag-metakolin based geopolymeric system. Ahmari et al. (2012) reported a comparable observation when they prepared geopolymer specimens with different proportions of ground waste concrete (GWC) powder. Their observations ascertained that the inclusion of GWC powder improved the strength of geopolymer binding because of the formation of C–S–H gel in the geopolymer system. However, other studies did not locate C–S–H gel in the geopolymeric system without OPC additions; for example, Rickard et al. (2012) did not report the formation of C–S–H phases when they studied the phase composition of geopolymers synthesised from five different fly ashes gleaned from five separate locations, namely Collie, Port Augusta, Eraring, Bayswater and Tarong. This indicates that OPC is a beneficial additive that can result in the precipitation of calcium silicate hydrate phases and at the same time improved the dissolution of fly-ash in the alkaline medium and subsequently the geopolymerisation reaction.

Density and Porosity

The density and porosity of geopolymer matrices and CF-reinforced geopolymer composites are shown in Tables 3.7 and 3.8. The density of geopolymer matrices ranged from 1.85 to 2.21 g/cm^3 (see Table 3.7) and the density of CF-reinforced geopolymer composites ranged between 1.78 and 1.91 g/cm^3 (see Table 3.8). The porosity of geopolymer matrices ranged from 23.8 to 18.6%, whereas the porosity of geopolymer composites varied from 26.6 to 21.7% (see Table 3.8). The porosities of geopolymer matrices and composites are reduced by the addition of OPC. This indicates that OPC has a pore-filling effect in geopolymer paste composites with or without CF; this result agrees with the work done by Jo et al. (2007) where the porosity of cement mortar is decreased by the addition of nano-SiO_2 particles. This reduced porosity level can be attributed to an increased matrix density because of decalcification of the C–S–H gels, which fill the pore structure of the geopolymer paste and yield a more consolidated microstructure, as reported by Bernal et al. (2010). Figure 3.10 shows the microstructure of a geopolymer composite containing 10 wt% OPC and the SEM–EDS provides evidence of the formation of C–S–H within the geopolymeric gels. Pressing the top surface of the samples could also contribute to a reduction in porosity by expelling trapped air from inside the sample and forcing cement into the voids and pore spaces. These factors may have resulted in the reduction in porosity and an increase in density.

Table 3.7 Densities and porosities of geopolymer matrices

Sample	OPC content (wt%)	Density (g/cm^3)	Porosity (%)
GP	0	1.85 ± 0.03	23.8
GP/OPC5	5	1.96 ± 0.02	21.7
GP/OPC8	8	2.18 ± 0.05	19.2
GP/OPC10	10	2.21 ± 0.03	18.6

Table 3.8 Densities and porosities of CF-reinforced geopolymer composites

Sample	OPC content (wt%)	Density (g/cm^3)	Porosity (%)
GP/CF	0	1.78 ± 0.02	26.6
GP/CF/OPC5	5	1.84 ± 0.04	24.5
GP/CF/OPC8	8	1.89 ± 0.02	22.5
GP/CF/OPC10	10	1.91 ± 0.03	21.7

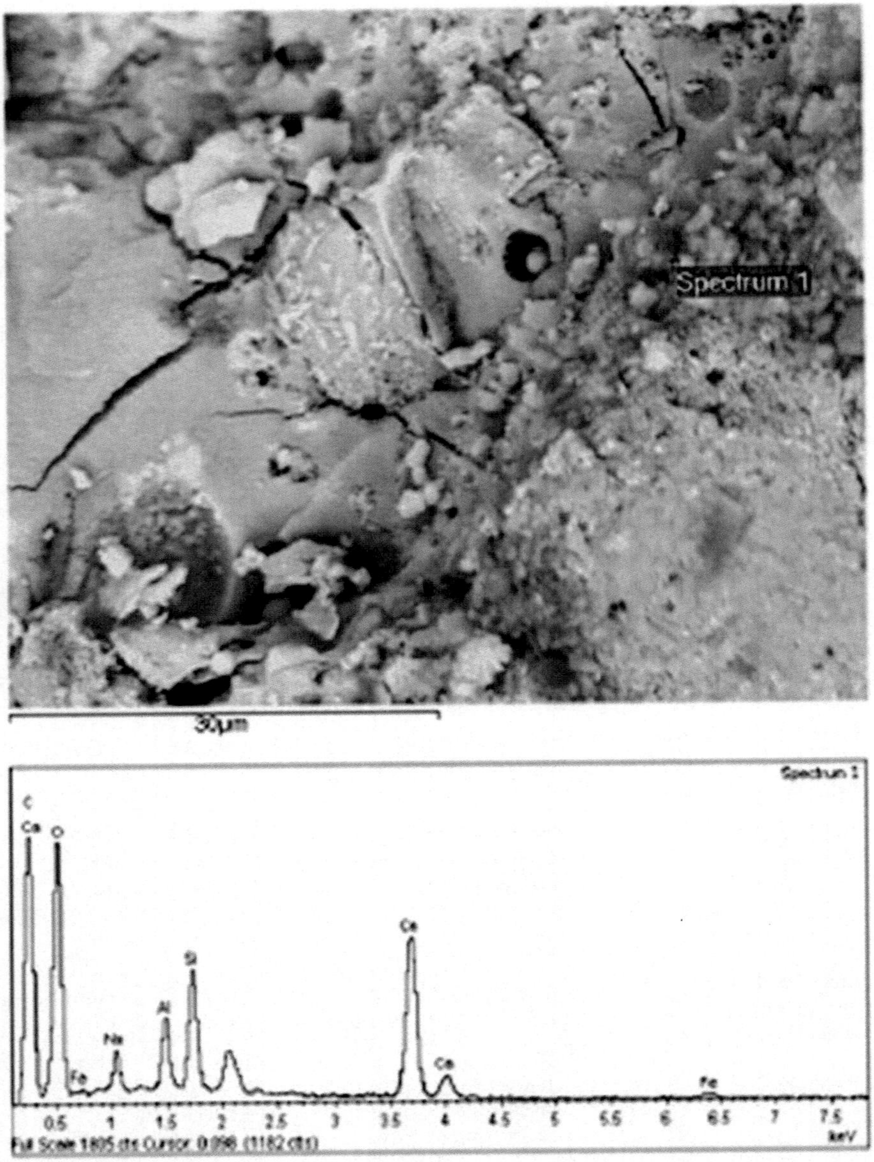

Fig. 3.10 SEM micrograph and EDS of the fracture surface region of geopolymer composite containing 10 wt% OPC

3.3 Flax Fabric-Reinforced Geopolymer Composites

3.3.1 Effect of Flax-Fibre Content

The compositions of the fabricated samples are shown in Table 3.9. The morphology of the flax fabric is shown in Fig. 3.11.

FTIR Observation

FTIR spectra of both pure geopolymer and flax/geopolymer composite are shown in Fig. 3.12. The strong peak at ~1000 cm^{-1} is associated with Al–O and Si–O asymmetric stretching vibrations and is the fingerprint of the geopolymerisation (Phair and Van Deventer 2002). The FTIR spectra show a broad peak in the region at 3466 cm^{-1} corresponding to the hydroxyl (OH) stretching vibration of free and hydrogen bonded –OH groups (Karbowiak et al. 2011; Lasagabaster et al. 2009), and the absorbance peak around 1653 cm^{-1} is attributed to the bending vibration of absorbed water (Zaharaki et al. 2010; Tserki et al. 2005). The presence of bands in the regions 1440–1490 cm^{-1} is an indicator of the atmospheric carbonation on the surface of the matrix where it reacts with carbon dioxide [34]. The presence of flax fibres in the composites can be recognised by the peak at 1418 cm^{-1}, which is attributed to the CH_3 bending vibration of cellulose (Karbowiak et al. 2011). The

Table 3.9 Formulation of samples

Sample	Fly ash (g)	NaOH solution (g)	Na_2SiO_3 solution (g)	Fabric layers	FF content (wt%)
1	1000	214.5	535.5	0	0
2	1000	214.5	535.5	5	2.4
3	1000	214.5	535.5	7	3.0
4	1000	214.5	535.5	10	4.1

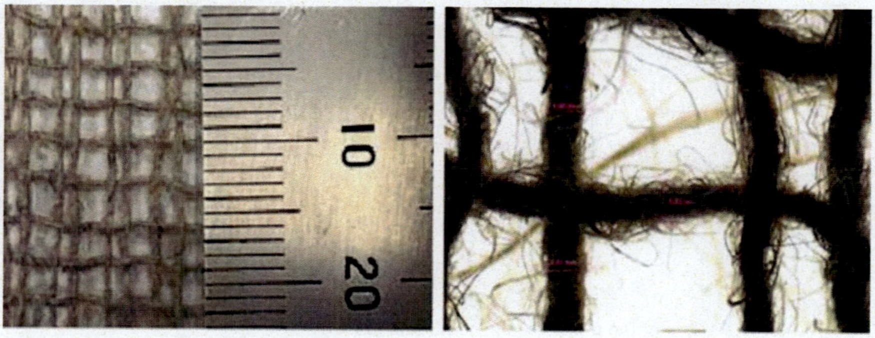

Fig. 3.11 Structure of the flax fabric

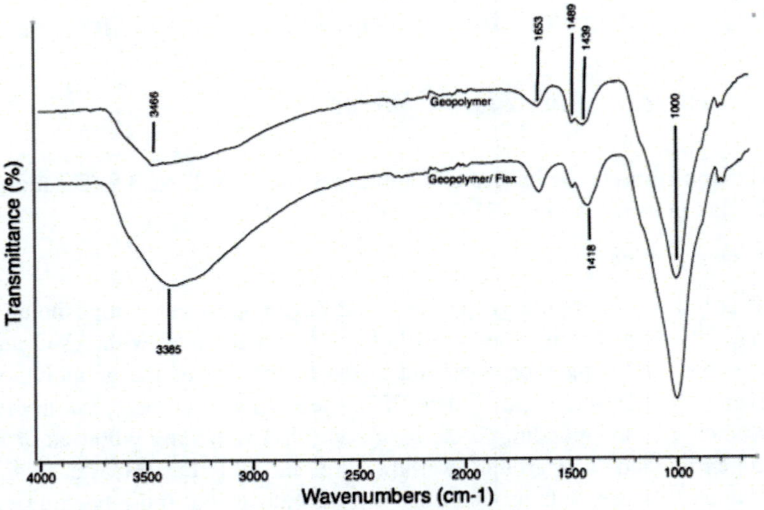

Fig. 3.12 FTIR spectra of pure geopolymer and geopolymer/flax composites

intensity of the band at 3385 and 1653 cm^{-1} increases in response to the existence of absorbed water in the cellulose fibres.

3.3.2 *Effect of Nanoclay and Flax-Fibre Content*

The structure, morphology and physical properties of the flax fabric are shown in Fig. 3.13 and Table 3.10. The compositions of the fabricated samples are indicated in Table 3.11.

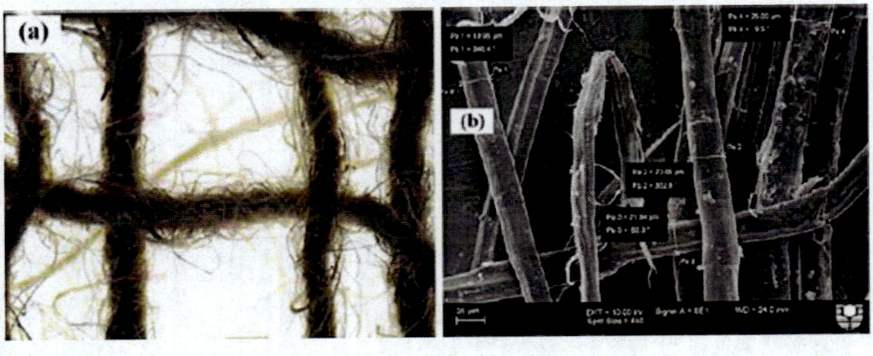

Fig. 3.13 Diameters of the **a** flax bundle and **b** flax fibres

Table 3.10 Structure and physical properties of the flax fabric

Fabric thickness (mm)	0.6
Fabric geometry	Woven (plain weave)
Yarn nature	Bundle
Bundle diameter (mm)	0.6
Filament size (mm)	0.01–0.02
Opening size (mm)	2–4
Fabric density (g/cm^3)	1.5
Modulus of elasticity (GPa)	39.5
Tensile strength (MPa)	660

Table 3.11 Formulation of samples

Sample name	Fly-ash (g)	NaOH solution (g)	Na_2SiO_3 solution (g)	Nanoclay (g)	FF content (wt%)
GP	1000	214.5	535.5	0	0
GPNC-1	1000	214.5	535.5	10	0
GPNC-2	1000	214.5	535.5	20	0
GPNC-3	1000	214.5	535.5	30	0
GPFNC-0	1000	214.5	535.5	0	4.1
GPFNC-1	1000	214.5	535.5	10	4.1
GPFNC-2	1000	214.5	535.5	20	4.1
GPFNC-3	1000	214.5	535.5	30	4.1

Physical Properties

Figure 3.14 shows the values of porosity and water absorption of all fabricated samples. It can be seen in general that the composites containing FF have higher porosity and water absorption than those composites without FF. This is because of the hydrophilic nature of cellulose fibres, which creates voids in the interfacial region between the flax fibres and the matrices (Alomayri et al. 2014).

All geopolymer nanocomposites displayed higher densities and lower porosities than the control paste. This indicates that nanoclay particles played a pore-filling role to reduce the porosity of the geopolymer composites, producing dense geopolymer paste. Because of this, the geopolymer nanocomposites exhibited lower water absorption. The optimum addition was found as 2.0 wt% of nanoclay, which reduced the porosity by 7.1%, and the water absorption by 17% when compared to the pure geopolymer matrix. However, the addition of excessive amounts of nanoclay increased the porosity and water absorption, and decreased the density of the nanocomposite sample due to the poor dispersion and agglomeration of nanoparticles (Alamri et al. 2012). This is a common phenomenon for nanoparticles due to small sizes, and high surface area to volume ratio of nanoparticles (van der Waal's force) (Shaikh et al. 2014). Images (a) and (b) in Fig. 3.15 show SEM images of

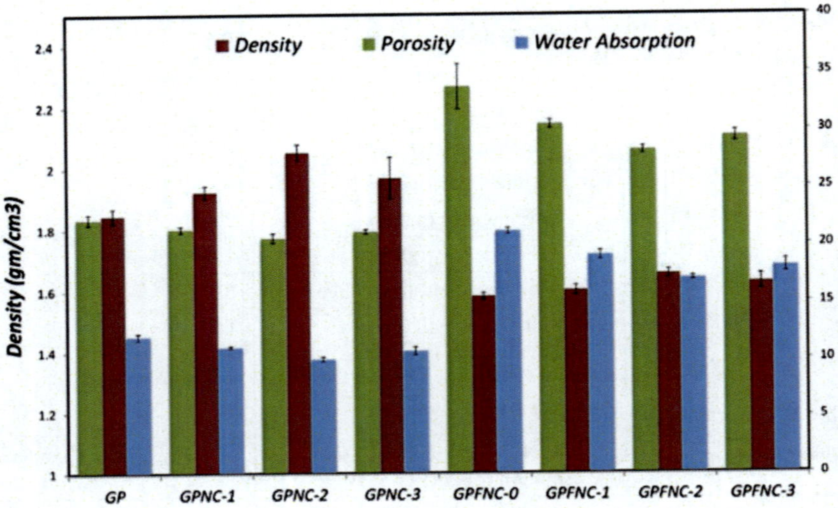

Fig. 3.14 Density, porosity and water absorption values for all samples

agglomerated nanoclay particles in GPNC-3 sample with Energy Dispersive Spectroscopy (EDS) spectra, ammonium salt in the nanoclay is identified by carbon and nitrogen elements. The nitrogen element is not detected clearly in the spectra because the nitrogen content is very low. However, the carbon content is clearly detected at 0.25 keV. This result is comparable with physical properties where the porosity of cement paste is decreased due to addition of 1.0 wt% of nanoclay to cement paste. Nevertheless, after the addition of more nanoclay to the paste, values of porosities and water absorption have increased because of the effect of nanoparticles agglomeration (Hakamy et al. 2014). SEM images (a) and (b) in Fig. 3.16 show microstructures of neat geopolymer and nanocomposites containing 1.0, 2.0, and 3.0 wt% nano-clay. The pure geopolymer matrix has a porous structure with a higher number of non-reacted and partially reacted fly ash particles embedded in the matrix (Fig. 3.15a). For the 1–3 wt% nano-clay (Fig. 3.16b–d) less fly ash particles were observed, and the matrix seemed denser when compared to the matrix of the control sample.

In the case of FF reinforced nanocomposites, the physical properties show similar trends to that of the nanocomposite trends. The optimum loading of nanoclay to the composites was found as 2.0 wt% in the case of GPFNC-2, which decreased the value of porosity by 16.3% and water absorption by 19.4% lower than the sample GPFNC-0.

X-Ray Diffraction (XRD)

The XRD spectra obtained for nanoclay, flax fibres, fly ash, GPNC-0 and GPNC-3 specimens are presented in Fig. 3.17a, b. The crystalline phases were indexed using Powder Diffraction Files (PDFs) from the Inorganic Crystal Structure Database (ISCD). Figure 3.17a shows the diffraction patterns of nanoclay and flax fibres. Three phases have been indexed in the diffraction pattern of nanoclay with the major phase

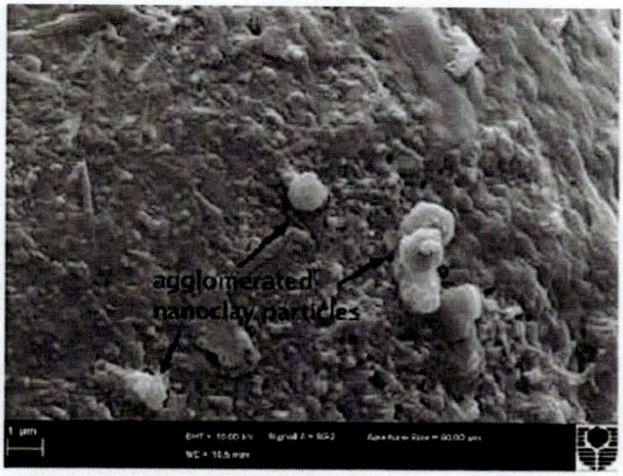

(a)

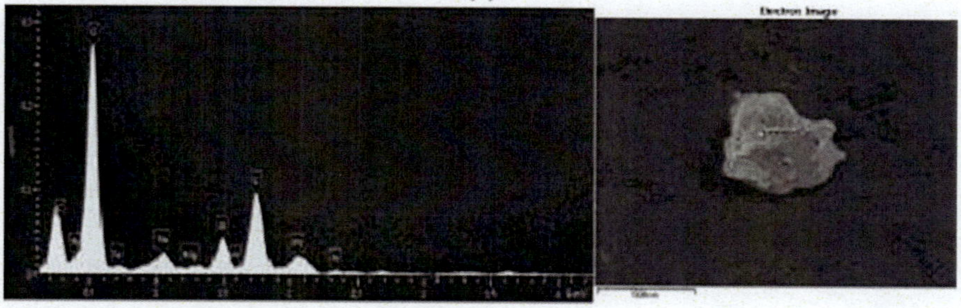

(b)

Fig. 3.15 **a** SEM image of agglomerated nanoclay particles on the fracture surface of GPNC-3, **b** with EDS analysis

being Cloisite 30B (Ebadi-Dehaghani et al. 2015), and minor phases of Cristobalite [SiO_2] (PDF 00-039-1425) and Quartz [SiO_2] (PDF 00-047-0718). Cloisite 30B consists of Montmorillonite [$(Ca,Na)_{0.3}Al_2(Si,Al)_4O_{10}(OH)_2{\cdot}xH_2O$] and the quaternary ammonium salt. Montmorillonite has four major peaks in the XRD pattern, which correspond to 2θ of 4.84°, 19.74°, 35.12° and 53.98°. The quaternary ammonium salt has four peaks at 2θ of 4.84°, 9.55°, 24.42° and 29.49°. Note that there is an overlap of peaks at 2θ of 4.84° for Montmorillonite and quaternary ammonium salt. Both Cristobalite and Quartz has a peak that corresponds to 2θ of 21.99° and 26.61°, respectively. The diffraction pattern of flax fibres shows typical peaks of cellulose (PDF 00-060-1502).

For fly ash, GP and GPNC-3 samples, two major phases are identified clearly: quartz [SiO_2] (PDF 00-046-1045) and mullite [$Al_{1.272}$ $Si_{0.278}$ $O_{4.864}$] (PDF 01-083-1881) (Fig. 3.17b). As the crystalline phases of quartz and mullite are also the fly ash phases they are insensitive to geopolymetric reactions, and their role is limited in geopolymer paste as filler particles (Fernández-Jiménez and Palomo 2005; Alomayri

Fig. 3.16 SEM images of the fracture surface of geopolymer nanocomposites with different loadings of nano-clay **a** pure geopolymer, **b** 1.0 wt%, **c** 2.0 wt% and **d** 3.0 wt%. Legend: 1. Pores and 2. Unreacted fly-ash

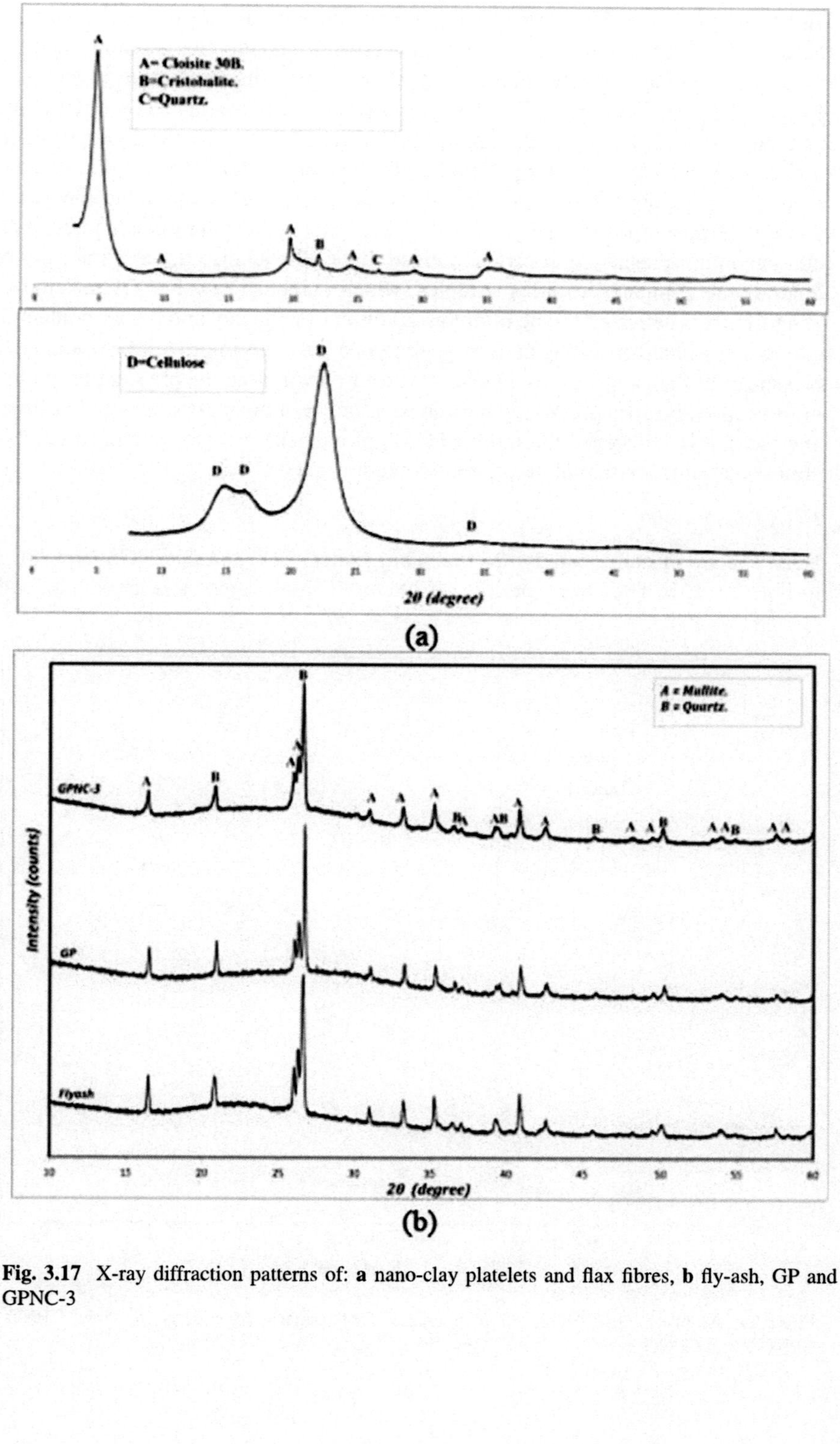

Fig. 3.17 X-ray diffraction patterns of: **a** nano-clay platelets and flax fibres, **b** fly-ash, GP and GPNC-3

and Low 2013a, b). However, the amorphous aluminosilicate phase that created between $2\theta = 14°$ and 27° is an active indication of geopolymer reaction, which is the reactive and dissolvable content in alkaline solution throughout the geopolymer formation (Chen-Tan et al. 2009). The geopolymer matrix mechanical properties are noticeably affected through the amorphous phase. When the amorphous phase is higher, the strength of the geopolymer is likewise higher (Bakharev 2006; Rickard et al. 2011). Figure 3.18 shows overlays of the amorphous hump under the quartz phase of nanocomposites samples. GPNC-2 has the highest amorphous phase over all nanocomposites. Also, it can be noticed that GPNC-2 displays less intensity of quartz peak compared to other samples, which demonstrates that the reaction of geopolymer is activated by the optimum addition of nanoclay and higher content of quartz is dissolved, resulting in more geopolymer gel. This improves the mechanical properties of the geopolymer nanocomposites by improving the physical properties of the matrix, besides improving the adhesion between the reinforcement flax fibres and the matrix. However, the more addition of nanoclay is inactive and resulted in almost the same amount of amorphous content as GPNC-1.

FTIR Observation

FTIR spectra of pure geopolymer, nanocomposites, GPF and GPFNC-2 are shown in Fig. 3.19a, b. The strong peak at ~1000 cm^{-1} in all samples is associated with

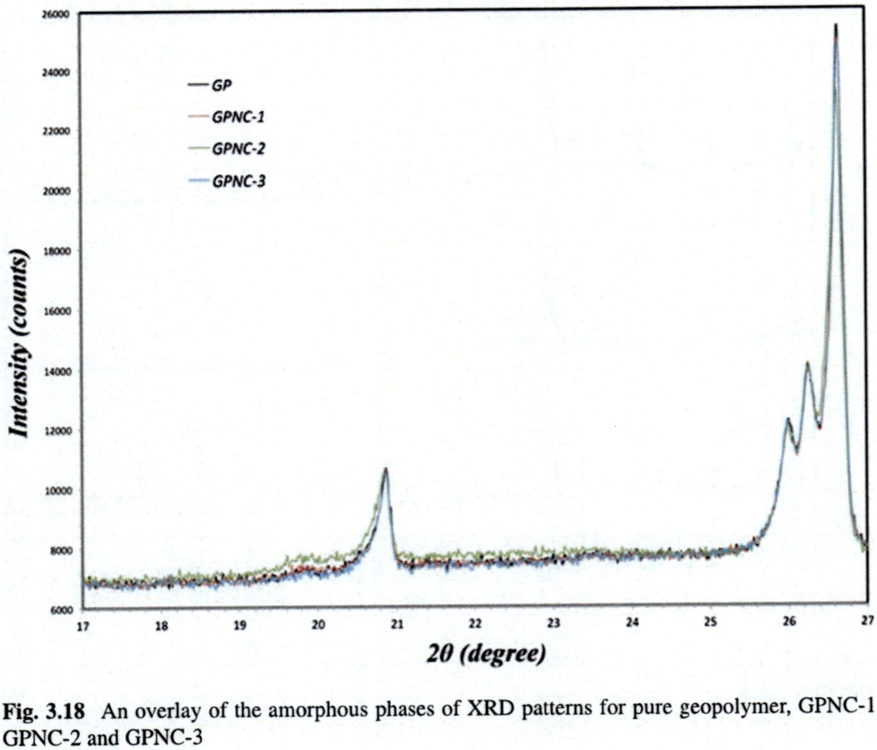

Fig. 3.18 An overlay of the amorphous phases of XRD patterns for pure geopolymer, GPNC-1, GPNC-2 and GPNC-3

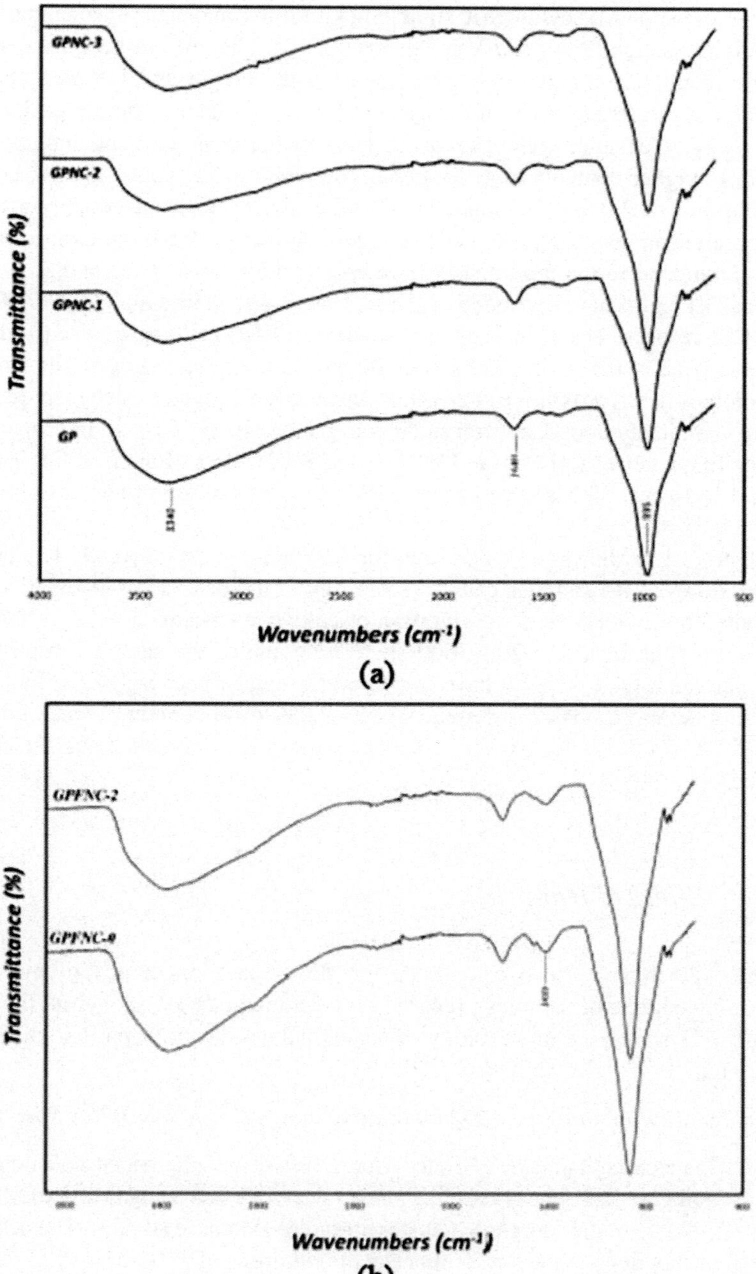

Fig. 3.19 **a** FTIR spectra of pure geopolymer and the nanocomposites GPNC-1, GPNC-2 and GPNC-3, **b** FTIR spectra of the FF-reinforced geopolymer composite GPFNC-0 and GPFNC-2

Si–O–T (T: Si or Al) asymmetric stretching vibrations and is the special mark of the geopolymerisation (Phair and Van Deventer 2002). The level of geopolymerization can be identified quantitatively by comparing the height and the area under the geopolymer stretching peaks of the nanocomposites to the pure matrix peak (Ul Haq et al. 2014). Considering the size of the geopolymer peak, all nanocomposites had generally higher contents of geopolymer compared to the control paste; however, the addition of 2.0 wt% of nanoclay had the highest level of geopolymerization among all samples. The areas under the geopolymer peak for the nanocomposites when compared to the pure matrix have enlarged by 2.0%, 7.0% and 3.0%, while the peak's heights have expanded by 2.0%, 19% and 15% for GPNC-1, GPNC-2 and GPNC-3, respectively. This result agrees with the XRD results that discussed above. A broad peak at the region of 3200–3600 cm^{-1} is corresponding to the stretching vibration of the hydroxyl (OH) group of physically free water (higher frequencies), and to chemically bounded water to the inorganic polymer through hydrogen bonds (lower frequencies) (Alamri and Low 2013; Kanny and Mohan 2014). The peak around 1640 cm^{-1} is also due to the (OH) bending vibration of absorbed water (Ul Haq et al. 2014).

Figure 3.19b shows the FTIR scan for GPFNC-0 and GPFNC-2. The presence of flax fibres in the samples can be recognised in the peak at 1420 cm^{-1}, which is attributed to the CH_2 bending vibration of cellulose (Alamri and Low 2013). The intensity of the band at 3200–3600 cm^{-1} is a sign to the samples water uptake. Samples reinforced with FF have higher water uptake because of the hydrophilic nature of cellulose fibres; however, GPFNC-2 has lower content of water compared to GPFNC-0 due to the barrier property of the nanocomposites against moisture uptake.

3.3.3 *Effect of Nanosilica*

Figure 3.20 depicts the two methods for the preparation of geopolymer pastes. The compositions of samples prepared by both methods are shown in Table 3.12. Figure 3.21 shows the morphology of nanosilica with average particle diameter of 18–25 nm.

X-Ray Diffraction Analysis and Quantitative Energy Dispersive X-Ray Spectrometer

The XRD spectra obtained for NS, flax fibres, fly ash and all nanocomposite samples are given in Fig. 3.22. The crystalline phases were indexed using Powder Diffraction Files (PDFs) from the Inorganic Crystal Structure Database (ICSD). The diffraction pattern of flax fibers shows typical peaks of cellulose (PDF 00-060-1502), whereas the NS powder displays a complete amorphous (glass) phase. In the case of fly ash, two main phases are indexed distinctly: quartz [SiO_2] (PDF 00-046-1045) and mullite [$Al_{2.32}Si_{0.68}O_{4.84}$] (PDF 04-016-1588). Quartz and mullite are the major crystalline phases of the Eraring fly-ash [7, 23]. Therefore, they are unreactive in the geopolymeric reaction, and act as filler in geopolymer matrices [21, 22]. Nevertheless, the

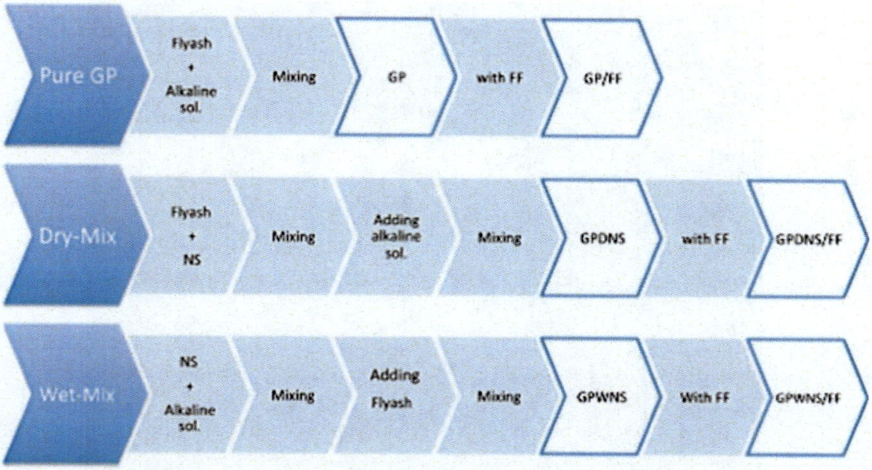

Fig. 3.20 Schematic showing the preparation of geopolymer; geopolymer nanocomposites and FF-reinforced geopolymer nanocomposites in dry/wet mix procedures

Table 3.12 Formulation and composition of samples prepared using dry and wet mix procedures

Sample	Fly-ash (g)	NaOH (g)	Na_2SiO_3 (g)	Water (g)	Nanosilica (g)
GP	1000	214.5	535.5	50	0
GPNS-0.5	1000	214.5	535.5	50	5
GPNS-1.0	1000	214.5	535.5	50	10
GPNS-2.0	1000	214.5	535.5	50	20
GPNS-3.0	1000	214.5	535.5	50	30

Mass ratios: (SiO_2/Na_2O) of sodium silicate = 2.0; (Na_2O/fly ash) = 0.079

amorphous aluminosilicate broad hump produced between $2\theta = 14°$ and 27° characterizes the reactive and dissolvable content in alkaline solution during the geopolymer development, which determines the activity of geopolymeric reaction (Ul Haq et al. 2014; Jun and Oh 2015; Ferone et al. 2013). The degree of amorphous fly ash is recognized as one of the most important factors that influence the physical and mechanical properties of fly ash geopolymers: the higher the amount of amorphous phase, the greater the strength exhibited by the geopolymer (Ul Haq et al. 2014; Jun and Oh 2015). Plots in Fig. 3.22b, c show the XRD spectra of pure geopolymer and geopolymer nanocomposites prepared by dry and wet mix preparation procedures. All samples have the main crystalline phases, quartz and Mullite, besides the fluorite [CaF_2] (PDF 04-002-2191). Fluorite is the standard used to determine the weight percentage of each crystalline phase. The phase abundance of crystalline phases in each sample was determined using Rietveld refinement, and the amorphous contents were calculated. Plots in Fig. 3.23 show Rietveld refinement of the diffraction pattern of pure geopolymer and GPDNS-1. In general, the addition of 0.5, 1.0,

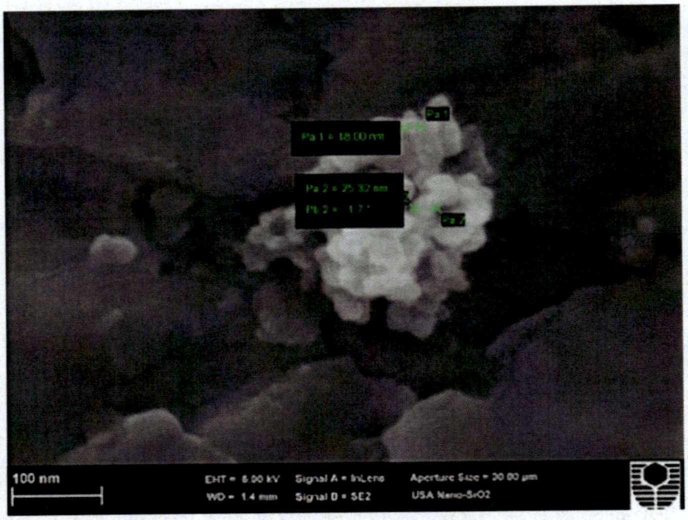

Fig. 3.21 SEM image of nano-silica particles

2.0 and 3.0 wt% NS into the geopolymer matrix has resulted in slight changes to the crystalline and amorphous contents in the samples. As can be seen from Fig. 3.24, the addition of 3.0 wt% NS increased the amorphous content in both dry and wet mix samples by 3.2% and 2.5%, respectively compared to the pure geopolymer paste. Because of this, the relative amounts of crystalline phases for the dry and wet mix nanocomposites were reduced. The growth of the amorphous content in the nanocomposite samples could be attributed to two reasons. First, the amorphous nature of the unreacted NS contributes to the total amorphous contents in the nanocomposites. This effect is believed to occur more in the case of dry-mix samples since a part of the NS particles perform as a filler in the matrices. Secondly, the addition of NS to the system promotes geopolymeric reaction producing higher amorphous amount of geopolymer gel in the nanocomposite.

The silicon-to-aluminum (Si/Al) ratio has a significant impact on the physical and mechanical properties of geopolymers (Rowles and O'Connor 2003; Duxson et al. 2007). Table 3.13 shows the Si/Al ratio as prepared and calculated theoretically in all samples, and the Si/Al ratios as determined experimentally using QEDS analysis. The theoretical ratios are similar in dry and wet mix nanocomposites as the amounts of silica added to the system are the same in both cases. However, preparation and mixing approaches could control the way the nanoparticles disperse in matrices, which produce samples with different ratios depending on the mixing methods. EDS analysis was conducted to determine the experimental ratios of Si/Al. EDS investigation was restricted to just the geopolymer gel areas in both dry and wet mix samples. Five spots at different location from the geopolymer gel are detected and averaged. Generally, the Si/Al ratios increased with the addition of NS particles in all samples due to the silica added to the system. The ratio started from 2.29%

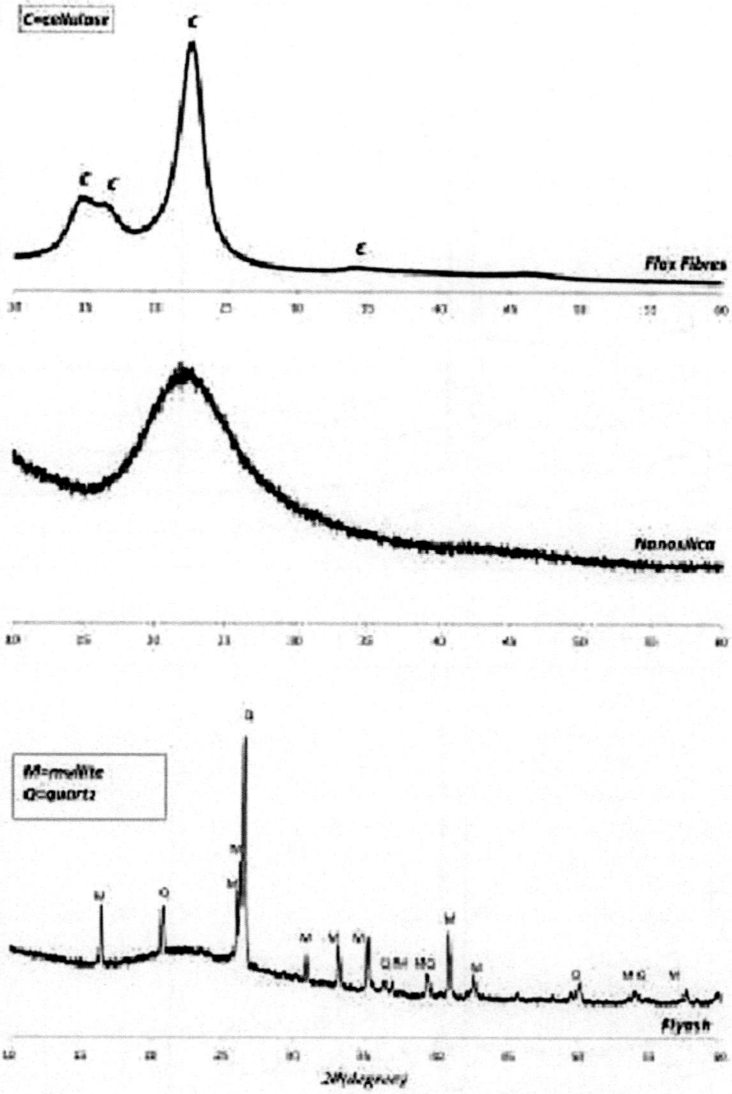

(a)

Fig. 3.22 XRD patterns of: **a** flax fibres, nanosilica and fly ash. **b** Geopolymer nanocomposites prepared by dry-mix procedure. **c** Geopolymer nanocomposites prepared by wet-mix procedure. Letters indicate: M = mullite, Q = quartz and F = fluorite

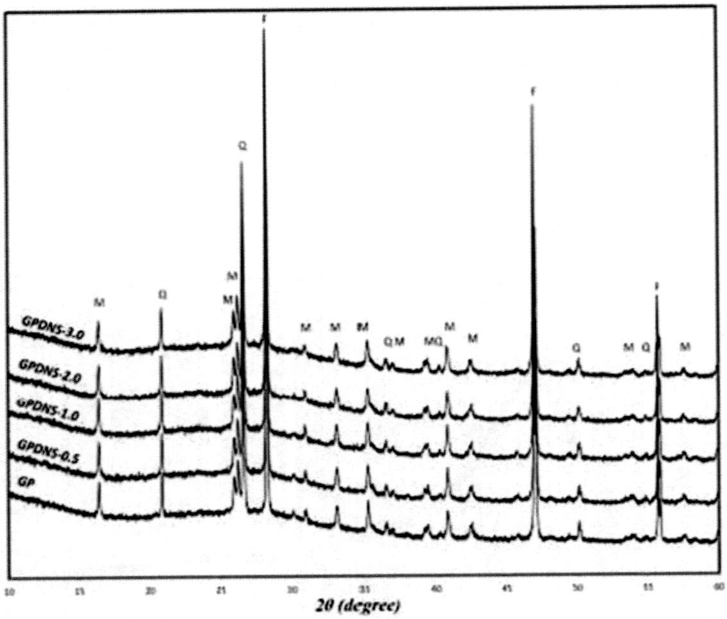

(b)

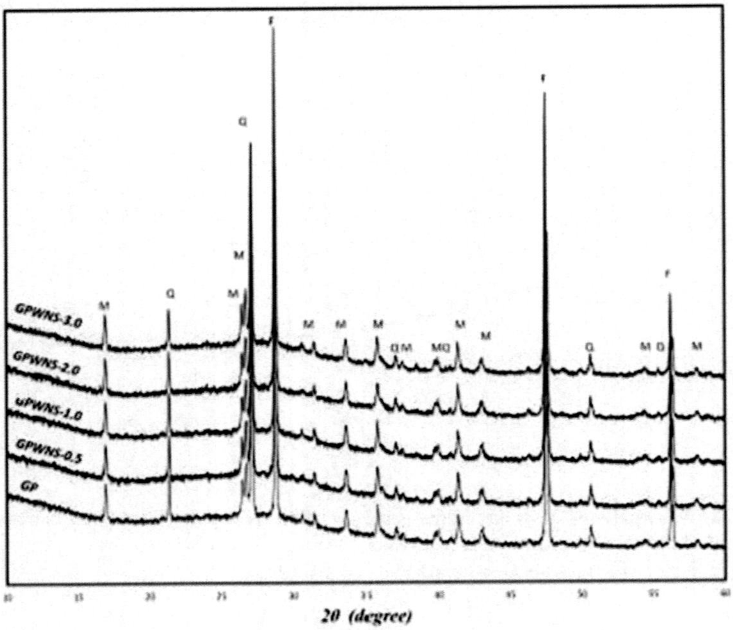

(c)

Fig. 3.22 (continued)

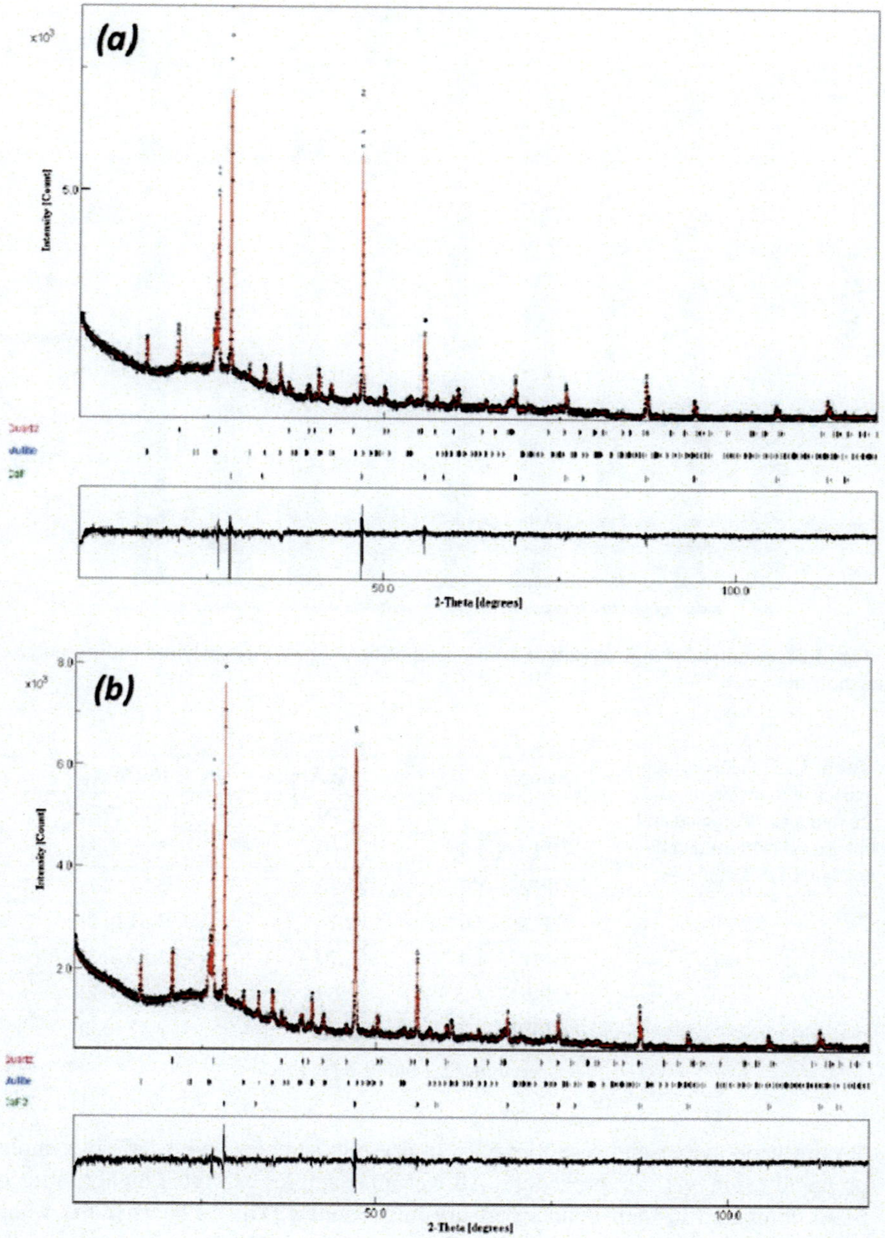

Fig. 3.23 XRD Rietveld plots for: **a** GP; and **b** GPDNS-1. Measured patterns are indicated by black points, and calculated patterns by solid red lines. The green residual plot shows the difference between the calculated and the measured pattern

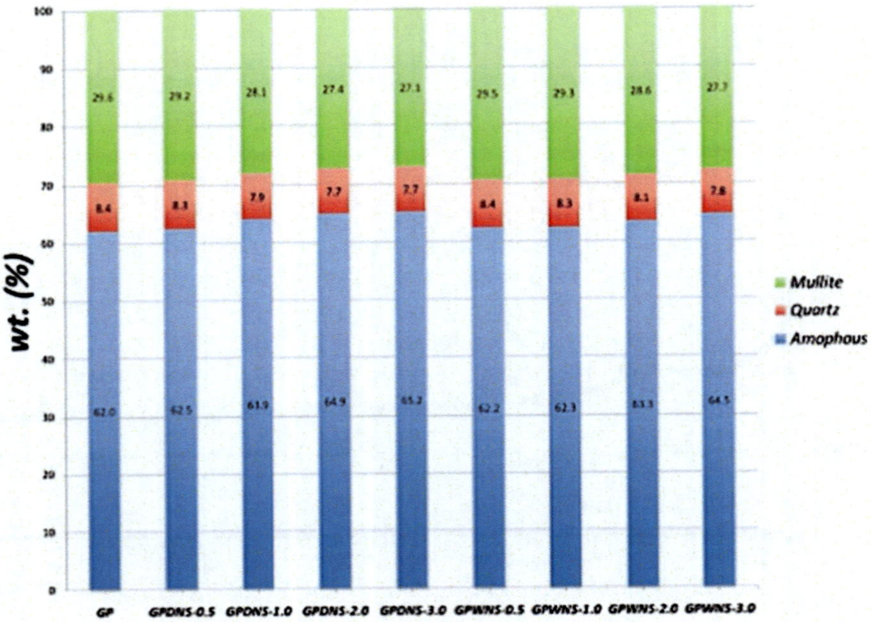

Fig. 3.24 Amorphous and crystalline phase compositions of pure geopolymer and geopolymer nanocomposites

Table 3.13 Geopolymer pastes with different silica contents as calculated and measured experimentally using QEDS technique

Sample	$(Si:Al)_{calc}$	$(Si:Al)_{exp}$
GP	2.69	2.29 (0.2)
GPDNS-0.5	2.69	2.66 (0.3)
GPDNS-1.0	2.72	3.02 (0.2)
GPDNS-2.0	2.76	3.41 (0.2)
GPDNS-3.0	2.79	3.57 (0.3)
GPWNS-0.5	2.69	2.75 (0.2)
GPWNS-1.0	2.72	3.63 (0.2)

Uncertainties are indicated in brackets

at pure geopolymer and rose to 3.57% in dry-mix samples and 4.10% in samples prepared using wet-mix procedure. All wet-mix samples exhibited higher ratios of Si/Al compared to their counterpart dry-mix samples. This is because the whole amount of silica particles dissolved in the case of wet-mix approach increasing the silicon contents of geopolymer pastes, while in the dry-mix preparation method a part of the NS particles did not dissolve and played a filler role in geopolymer matrices. SEM images in Fig. 3.25a–c show the microstructures of pure geopolymer and geopolymer nanocomposites containing 3.0 wt% NS in wet and dry mix samples, respectively. High amount of unreacted and partially reacted fly ash particles can be

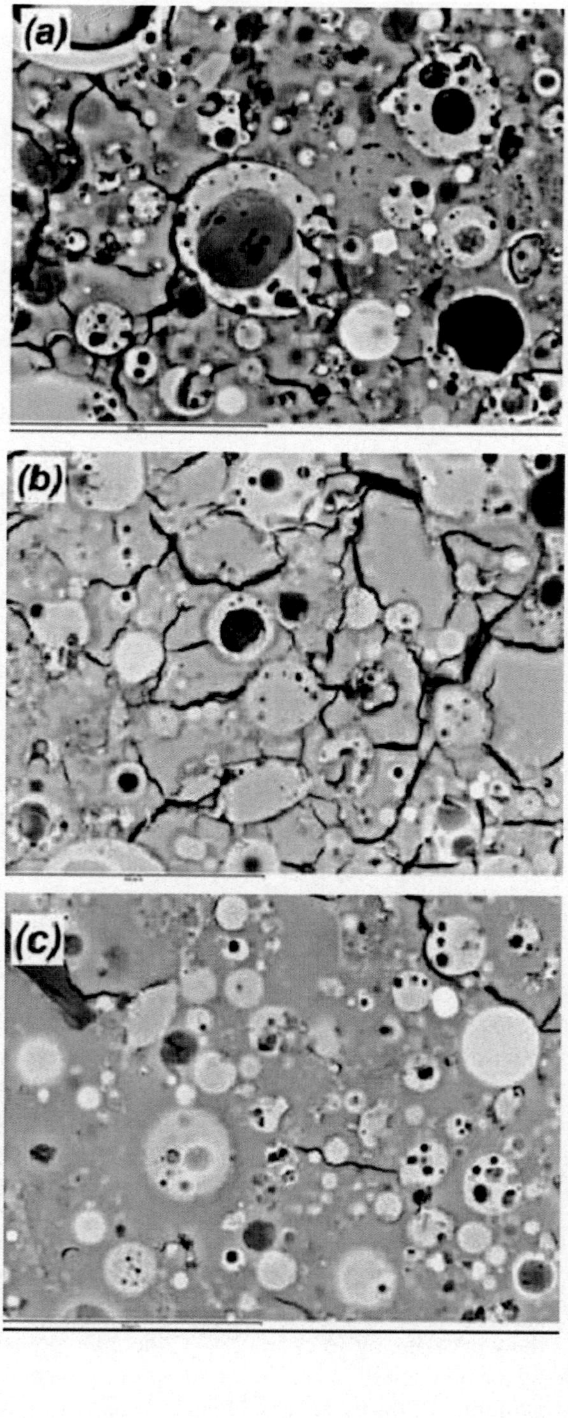

Fig. 3.25 SEM images show **a** pure geopolymer. Geopolymer nanocomposites containing 3.0 wt% NS prepared by: **b** wet-mix procedure, **c** dry mix procedure

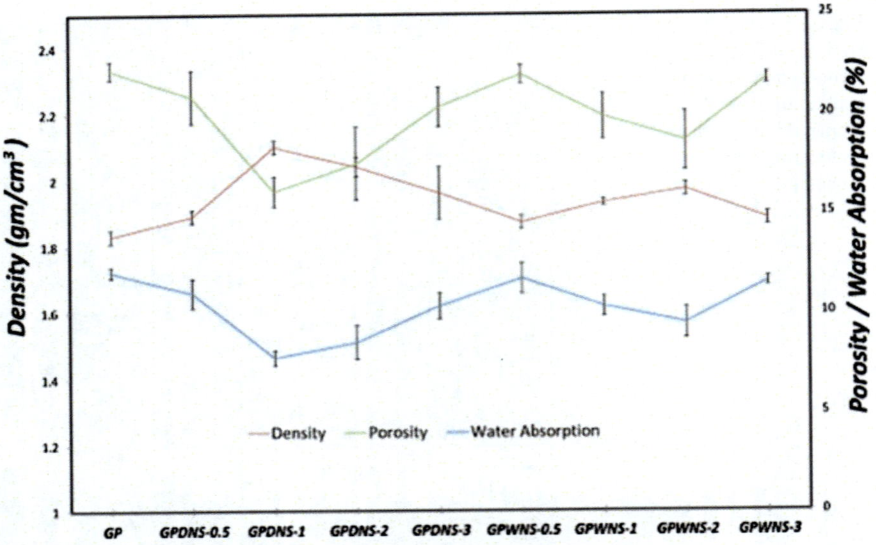

Fig. 3.26 Density, porosity and water absorption of pure geopolymer and geopolymer nanocomposites

clearly seen in the case of pure geopolymer. However, after the addition of NS higher amount of geopolymer gel, and fewer amount of unreacted fly ash particles appeared in the wet and dry mix samples.

Physical Properties

The density, porosity and water absorption of geopolymer paste and that containing NS are shown in Fig. 3.26. It is observed that the addition of NS decreases the porosity and water absorption of all geopolymer nanocomposites when compared to control geopolymer paste. In the wet-mix samples, the density is improved by 7.6%, while the porosity and water absorption are decreased by 16.2% and 21.5%, respectively. However, in the case of dry mix preparation method, the optimum loading was found as 1.0 wt% of NS, which improved the density by 15%, and reduced the porosity and water absorption by 27% and 35%, respectively, when compared to the control paste.

This development could be attributed to two factors. First, the addition of amorphous silica to the system in the form of NS particles has enhanced the geopolymeric reaction producing more geopolymer gel and denser matrices (Phoo-ngernkham et al. 2014). Secondly, the NS particles acted as pore-filler reducing the porosity of all dry-mix samples. This enhancement reveals that geopolymer nanocomposite with the optimum supplement of NS synthesized using dry and wet mix preparation methods yields consolidated dense microstructure. The current results are in agreement with the work done by Jo et al. (2007) where the porosity of cement mortar is reduced by the addition of NS particles. In another study, Supit and Shaikh reported that the addition of 2.0 wt% NS notably reduced the porosity of high volume fly ash concrete

(Supit and Shaikh 2014). However, the further addition of NS leads to an increase in porosity and water absorption and a drop-in density. This could be attributed to the poor dispersion and agglomerations of the high NS contents, which creates more voids in the matrix (Assaedi et al. 2016; Senff et al. 2013). Agglomeration and poor dispersion are common phenomena in nanoparticles. The high ratio of surface area to the volumes of the nanoparticles increases the adhesion forces between the particles resulting in agglomerated nanoparticles.

Acknowledgements The authors would like to thank Ms E. Miller from Applied Physics at Curtin University for assistance with SEM. The authors would like to acknowledge Dr. W. Rickard and Mr. L. Vickers for assisting in mechanical tests. The collection of diffraction data was funded by the Australian Synchrotron (PD 5341).

References

Abdullah A, Jamaludin SB, Noor MM, Hussin K (2011) Composite cement reinforced coconut fibre: physical and mechanical properties and fracture behavior. Aust J Basic Appl Sci 5:1228–1240

Aggarwal LK (1995) Bagasse-reinforced cement composites. Cem Concr Compos 17:107–112

Ahmari S, Ren X, Toufigh V, Zhang L (2012) Production of geopolymeric binder from blended waste concrete powder and fly ash. Constr Build Mater 35:718–729

Alamri H, Low IM (2013) Effect of water absorption on the mechanical properties of nanoclay filled recycled cellulose fibre reinforced epoxy hybrid nanocomposites. Compos Part A 44:23–31

Alamri H, Low IM, Alothman Z (2012) Mechanical, thermal and microstructural characteristics of cellulose fibre reinforced epoxy/organoclay nanocomposites. Compos Part B 43:2762–2771

Alomayri T, Low IM (2013a) J Asian Ceram Soc 30:223–230

Alomayri T, Low IM (2013b) Synthesis and characterization of mechanical properties in cotton fiber-reinforced geopolymer composites. J Asian Ceram Soc 1:30–34

Alomayri T, Assaedi H, Shaikh FUA, Low IM (2014) Effect of water absorption on the mechanical properties of cotton fabric-reinforced geopolymer composites. J Asian Ceram Soc 2:223–230

Assaedi H, Shaikh FUA, Low IM (2016) Effect of nano-clay on mechanical and thermal properties of geopolymer. J Asian Ceram Soc 4:19–28

Bakharev T (2006) Thermal behaviour of geopolymers prepared using class F fly ash and elevated temperature curing. Cem Concr Res 36:1134–1147

Bernal SA, de Gutierrez RM, Provis JL, Rose V (2010) Effect of silicate modulus and metakaolin incorporation on the carbonation of alkali silicate-activated slags. Cem Concr Res 40:8989–9007

Buchwald A, Hilbig H, Kaps CH (2007) Alkali-activated metakaolin-slag blends performance and structure in dependence of their composition. J Mater Sci 42:3024–3032

Chen-Tan NW, Van Riessen A, Ly CV, Southam DC (2009) Determining the reactivity of a fly ash for production of geopolymer. J Am Ceram Soc 92:881–887

Criado M, Fernández Jiménez A, Sobrados I, Palomo A, Sanz J (2012) Effect of relative humidity on the reaction products of alkali activated fly ash. J Eur Ceram Soc 32:2799–2807

Duxson P, Mallicoat SW, Lukey GC, Kriven WM, van Deventer JSJ (2007) The effect of alkali and Si/Al ratio on the development of mechanical properties of metakaolin-based geopolymers. Colloids Surf A 292:8–20

Ebadi-Dehaghani H, Khonakdar HA, Barikani M, Jafari SH (2015) Experimental and theoretical analyses of mechanical properties of PP/PLA/clay nanocomposites. Compos Part B 69:133–144

Fernández-Jiménez A, Palomo A (2003) Characterisation of fly ashes. Potential reactivity as alkaline cements. Fuel 82:2259–2265

Fernández-Jiménez A, Palomo A (2005) Composition and microstructure of alkali activated fly ash binder: effect of the activator. Cem Concr Res 35:1984–1992
Ferone C, Colangelo F, Roviello G, Asprone D, Menna C, Balsamo A et al (2013) Application-oriented chemical optimization of a metakaolin based geopolymer. Materials 6:1920–1939
Hakamy A, Shaikh FUA, Low IM (2013) Microstructures and mechanical properties of hemp fabric reinforced organoclay–cement nanocomposites. Constr Build Mater 49:298–307
Hakamy A, Shaikh FUA, Low IM (2014) Thermal and mechanical properties of hemp fabric-reinforced nanoclay–cement nanocomposites. J Mater Sci 49:1684–1694
Jo B-W, Kim C-H, Tae G-H, Park J-B (2007) Characteristics of cement mortar with nano-SiO_2 particles. Constr Build Mater 21:1351–1355
Jun Y, Oh J (2015) Use of gypsum as a preventive measure for strength deterioration during curing in class F fly ash geopolymer system. Materials 8:3053–3067
Kanny K, Mohan TP (2014) Resin infusion analysis of nanoclay filled glass fiber laminates. Compos Part B 58:328–334
Karbowiak T, Ferret E, Debeaufort F et al (2011) Investigation of water transfer across thin layer biopolymer films by infrared spectroscopy. J Membr Sci 370:82–90
Kouamo HT, Mbey JA, Elimbi A, Kenne Diffo BB, Njopwouo D (2012) Synthesis of volcanic ash-based geopolymer mortars by fusion method: effects of adding metakaolin to fused volcanic ash. Ceram Int 12:8842–8862
Lasagabaster A, Abad MJ, Barral L et al (2009) Application of FTIR spectroscopy to determine transport properties and water-polymer interactions polypropylene (PP)/poly(ethylene-co-vinyl alcohol) (EVOH) blend films: effect of poly(ethylene-co-vinyl alcohol) content and water activity. Polymer 50:2981–2989
Neithalath N, Weiss J, Olek J (2004) Acoustic performance and damping behavior of cellulose–cement composites. Cem Concr Compos 26:359–370
Phair JW, Van Deventer JSJ (2002) Effect of the silicate activator pH on the microstructural characteristics of wasre-based geopolymers. Int J Miner Process 66:121–143
Phoo-ngernkham T, Chindaprasirt P, Sata V, Hanjitsuwan S, Hatanaka S (2014) The effect of adding nano-SiO_2 and nano-Al_2O_3 on properties of high calcium fly ash geopolymer cured at ambient temperature. Mater Des 55:58–65
Rattanasak U, Chindaprasirt P (2009) Influence of NaOH solution on the synthesis of fly ash geopolymer. Miner Eng 22:1073–1078
Rickard WDA, Williams R, Temuujin J, Van Riessen A (2011) Assessing the suitability of three Australian fly ashes as an aluminosilicate source for geopolymers in high temperature applications. Mater Sci Eng A 528:3390–3397
Rickard WDA, Temuujin J, van Riessen A (2012) Thermal analysis of geopolymer pastes synthesised from five fly ashes of variable composition. J Non-Cryst Solids 358:1830–1839
Rowles M, O'Connor B (2003) Chemical optimisation of the compressive strength of aluminosilicate geopolymers synthesised by sodium silicate activation of metakaolinite. J Mater Chem 13:1161–1165
Senff L, Tobaldi DM, Lucas S, Hotza D, Ferreira VM, Labrincha JA (2013) Formulation of mortars with nano-SiO_2 and nano-TiO_2 for degradation of pollutants in buildings. Compos B Eng 44:40–47
Shaikh F, Supit S, Sarker P (2014) A study on the effect of nano silica on compressive strength of high volume fly ash mortars and concretes. Mater Des 60:433–442
Supit SWM, Shaikh FUA (2014) Durability properties of high-volume fly ash concrete containing nano-silica. Mater Struct 48:2431–2445
Tserki V, Zafeiropoulos NE, Simon F et al (2005) A study of the effect of acetylation and propionylation surface treatments on natural fibres. Compos Part A 36:1110–1118
Ul Haq E, Kunjalukkal Padmanabhan S, Licciulli A (2014) Synthesis and characteristics of fly ash and bottom ash based geopolymers—a comparative study. Ceram Int 40:2965–2971
Xu H, Van Deventer JSJ (2000) The geopolymerisation of alumino-silicate minerals. J Miner Process 59:247–266

Xu H, Van Deventer JSJ (2002) Geopolymerisation of multiple minerals. Miner Eng 15:1131–1139
Yip CK, van Deventer JSJ (2003) Microanalysis of calcium silicate hydrate gel formed within a geopolymeric binder. J Mater Sci 38:3851–3860
Yip CK, Lukey GC, Van Deventer JSJ (2005) The coexistence of geopolymeric gel and calcium silicate hydrates at the early stage of alkaline activation. Cem Concr Res 35:1688–1697
Zaharaki D, Komnitsas K, Perdikatsis V (2010) Use of analytical techniques for identification of inorganic polymer gel composition. J Mater Sci 45:2715–2724

Chapter 4
Mechanical Properties

Abstract This chapter describes the mechanical properties of geopolymer reinforced with natural fibres and/or nanofillers. Results show that the appropriate addition of natural fibres and/or nanofillers can improve the mechanical and fracture properties of geopolymer composites and nanocomposites. Results show that the mechanical properties of these materials are improved by increasing the content of natural fibres.

4.1 Cotton Fibre-Reinforced Geopolymer Composites

4.1.1 Flexural Strength and Modulus

The effects of fibre content on the flexural strength and flexural modulus of cotton fibre-reinforced geopolymer composites are shown in Figs. 4.1 and 4.2, respectively. In Fig. 4.1, experimental results indicate that the flexural strength of composites increases initially with increasing cotton fibre content of up to 0.5 wt%, and then decreases thereafter. The enhancement in flexural strength may be ascribed to the good dispersion of cotton fibres throughout the matrix which helps to increase the interaction or adhesion at the matrix/cotton fibre interface. Hence, this permits the optimum operation of stress-transfer from the matrix to the cotton fibres, thus resulting in the improvement of strength properties. However, the flexural strength of composites decreases when fibre content increases to more than 0.5 wt% where a high content of cotton fibres inhibits the non-homogeneity within the matrix such that agglomerations are formed which degrade the interfacial adhesion between the fibre and the matrix. In addition, these agglomerations may act as stress concentrators to cause reductions in flexural strength (Talimi and Rizvi 2008).

It was observed that increasing the content of cotton fibre caused discernible increase in matrix viscosity, which in turn allowed residual air bubbles to be introduced either through mixing or by being trapped in the geopolymer during pouring into the mould. These conditions may be implicated in sample failure at relatively low

I.-M. Low et al., *Cotton and Flax Fibre-Reinforced Geopolymer Composites*, Composites Science and Technology,
https://doi.org/10.1007/978-981-16-2281-6_4

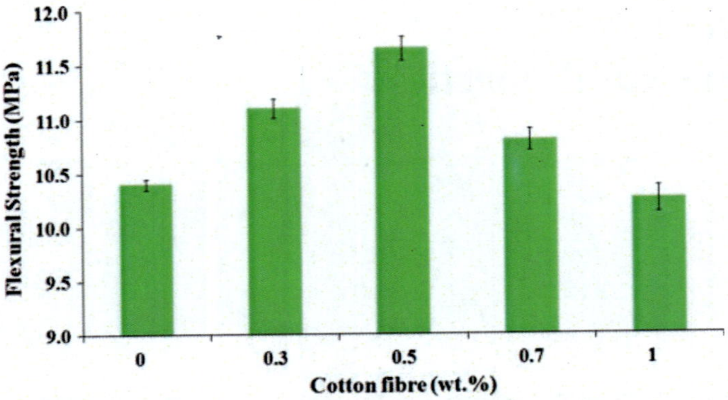

Fig. 4.1 Flexural strength of geopolymer composites as a function of fibre content

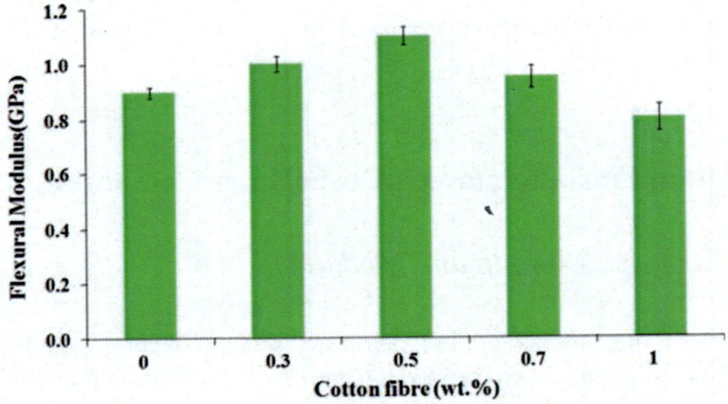

Fig. 4.2 Flexural modulus of geopolymer composites as a function of fibre content

stress. A lower loading of cotton fibres offers less potential for microvoid formation and more uniform dispersion; both contribute to strength improvement.

The flexural strength of the neat geopolymer paste increased from 10.4 to 11.7 MPa after the addition of 0.5 wt% cotton fibres. However, adding more cotton fibres (0.7 and 1.0 wt%) led to a reduction in strength.

The flexural moduli of geopolymer composites are shown in Fig. 4.2 and indicate similar trends to flexural strength values. The addition of 0.5 wt% cotton fibres in the geopolymer matrix increases the flexural modulus over plain geopolymer, but this trend reverses, reducing to 0.95 and 0.80 GPa, with the addition of 0.7 and 1.0 wt% cotton fibres. Two reasons may account for this observation: (1) increased viscosity, voids, and poor dispersion due to high cotton fibre content; and (2) presence of high proportion of other constituents (e.g. quartz and mullite) which act as inactive fillers and thus leads to insufficient geopolymer binders. The presence of quartz in a source material is particularly undesirable when designing geopolymers because it can cause microcracking, which reduces the strength of the material. This problem becomes more significant as the particle size of the quartz increases (Rickard et al. 2011). The presence of small amount of cotton fibres in the geopolymer matrix serves to counteract this, thereby increasing the flexural strength and flexural modulus of the geopolymer composites over plain geopolymer. The optimum content of cotton fibres in geopolymer composites is 0.5 wt%.

4.1.2 Fracture Toughness

The effect of cotton fibre content on the facture toughness of geopolymer composites is presented in Fig. 4.3. Cotton fibres play a significant role in enhancing the facture toughness of the matrices through several energy-absorbing functions such as fibre rupture, fibre/matrix interface debonding, fibre pull-out and fibre-bridging which slow crack propagation and therefore increase fracture energy (Reis 2006; Silva et al. 2009, 2010; Tolêdo Filho et al. 2000, 2003). The fracture toughness of geopolymer reinforced with 0.5 wt% cotton fibres increases by 1.12 MPa $m^{1/2}$ over neat geopolymer. This significant enhancement in facture toughness is due to fibre pull-out, fibre fracture and fibre-bridging, as clearly shown in the SEM images of Fig. 4.4b–e.

Some short fibres, such as poly vinyl alcohol (PVA) and basalt, have previously been employed to improve the mechanical performance of geopolymers because they provide some control of cracking and increase the fracture toughness of a brittle matrix by their bridging action during both micro and macro-cracking. It has been reported that short PVA fibres with an optimum volume fraction of 1.0% ameliorated the brittle properties of ash-based geopolymer (Zhang et al. 2006). Similarly, Dias and Thaumaturgo (2005) investigated fracture toughness of geopolymeric concretes reinforced with basalt fibres and found that geopolymeric concretes with 0.5–1.0 wt% basalt fibres showed higher fracture toughness than Portland cement concretes. In

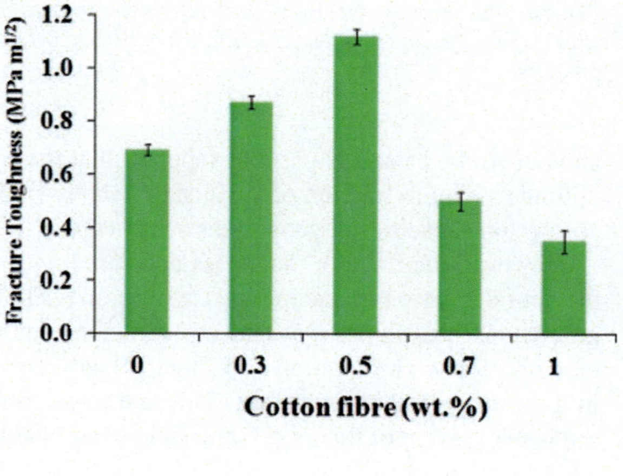

Fig. 4.3 Fracture toughness of geopolymer composites as a function of fibre content

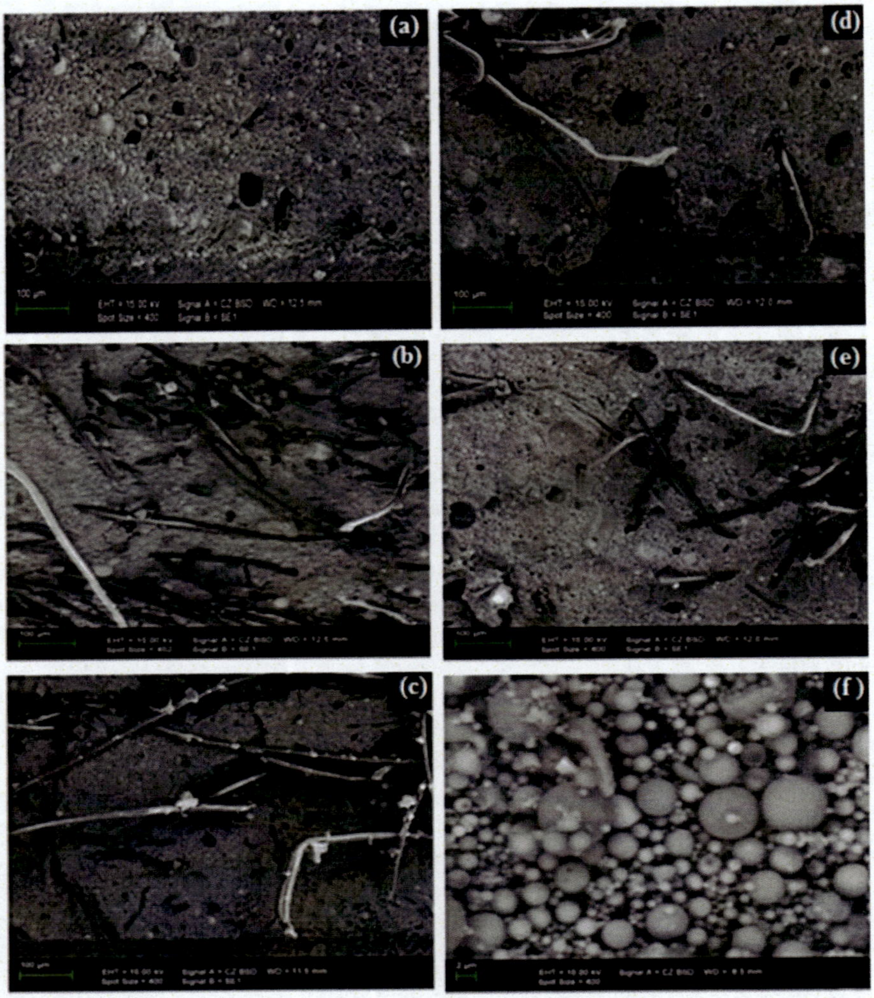

Fig. 4.4 SEM images of (f) fly ash and the fracture surface for geopolymer composites reinforced with varying content of cotton fibres **a** 0, **b** 0.3, **c** 0.5, **d** 0.7, **e** 1 wt%, and **f** morphology of fly-ash particles

another study, Li and Xu (2009) reported that the addition of basalt fibres with an optimum volume fraction of 0.3% significantly improved deformation and energy absorption capacities of geopolymeric concrete.

However, the fracture toughness decreased with increasing fibre content due to the poor dispersion of cotton fibres in the slurry. The dispersion of cotton fibre in the geopolymer matrix has a considerable influence on the properties of the fresh mix, on workability. The addition of 0.7 and 1.0 wt% cotton fibres resulted in a reduction in the consistency of the matrix. This had to be compensated for by an increase in the water content of the mix. Increasing water content to overcome such a problem may lead to other adverse effects, such as an increase in porosity and microcracking. These limitations usually lead to the reduction in bonding at the fibre-matrix interface, which results in lower stress transferred from the matrix to the fibres.

4.2 Cotton Fabric-Reinforced Geopolymer Composites

4.2.1 Effect of Fibre Content and Fabrication Methods

Two methods as described in Chap. 2 were used to prepare these composites. The first method involved a self-infiltration and hand lay-up technique. The second method used a forced-impregnation (wet out) and hand lay-up technique with a roller and brush.

(i) ***Self-infiltration and hand lay-up technique***

Table 4.1 shows the formulations of fabricated samples with different geopolymer binder and CF reinforcement percentages. These samples were used to investigate the effect of fibre content on the physical and mechanical properties of cotton fabric-reinforced geopolymer composites (CFG).

Density and porosity

Measured density and porosity of all composites are shown in Table 4.2. As the cotton fibre weight increased, the geopolymer composite density decreased. However, an increase in the fibre weight of composites caused a gradual increase in porosity.

Table 4.1 Formulations of samples

Sample	Fabric layers	Fabric mass (g)	Fibre content (wt%)
CFG0	0	0	0
CFG2	2	11.4	1.4
CFG3	3	17.2	2.1
CFG4	4	22.8	2.8
CFG6	6	34.2	4.1

Table 4.2 Density and porosity values of CFGs

Sample	Fibre content (wt%)	Density (g/cm^3)	Porosity (%)
CFG0	0	2.02 ± 0.03	21.1
CFG2	1.4	1.84 ± 0.02	24.8
CFG3	2.1	1.76 ± 0.02	26.2
CFG4	2.8	1.68 ± 0.03	29.6
CFG6	4.1	1.59 ± 0.05	32.6

This increase may be the result of voids becoming trapped beneath the CF sheets during casting, creating higher porosity and thus leading to poor adhesion between the fibre and matrix. These values closely agree with earlier experimental results by the authors (Alomayri et al. 2013a, b).

Flexural strength and flexural modulus

The effect of fibre content on the flexural strength of CFGs is shown in Fig. 4.5 and their corresponding stress/strain curves are shown in Fig. 4.6. It is observed that

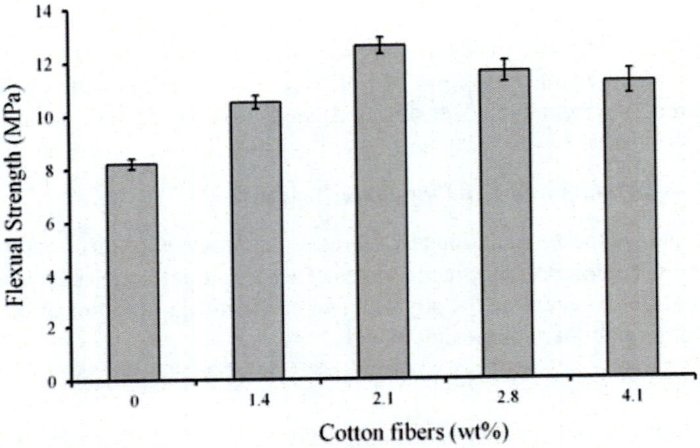

Fig. 4.5 Flexural strength of geopolymer composites as a function of fibre content

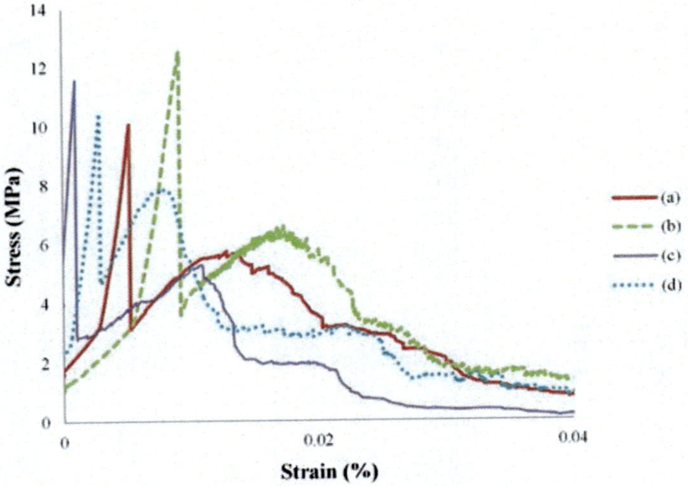

Fig. 4.6 Typical stress–strain curves of geopolymer composites with various cotton fibre content **a** 1.4 wt%, **b** 2.1 wt%, **c** 2.8 wt%, and **d** 4.1 wt%

the flexural strength of geopolymer composites increases with increase in cotton fabric contents up to 2.1 wt% and after that the strength of the composites decreases. This study finds that 2.1 wt% of cotton fibre provides the highest flexural strength. The highest value of flexural strength, exhibited by 2.1 wt% fibre content, may be explained by the orientation of the fibres within the matrix. At this stage, fibres achieve the maximum level of orientation within the matrix. This is because when the load is applied, the stress is uniformly distributed among the fibres (Khan et al. 2012). As a result, the flexural strength of the composites achieves its maximum value.

Increasing the amount of cotton fibres to 2.8 and 4.1 wt% reduces the flexural strength. This reduction might be caused by misalignment of the cotton fabric, which is due to the hand lay-up procedure. This imperfection affects the mechanical properties of the composites because the misalignment can lead to the inability of the fibre to support stress transferred from the geopolymer matrix and poor interfacial bonding between the fibre and the matrix (Soroushian et al. 2006). Therefore, the strength of geopolymer composites decreased with the increase in the cotton fabric content beyond 2.1 wt%.

The same reasons can explain a similar trend observed with the flexural modulus (see Fig. 4.7). This reduction in modulus could additionally be attributed to the fact that the cotton fibres are hydrophilic in nature and tend to absorb moisture from ambient air, which in turn, causes swelling of the fibre, forming voids and micro cracks at the fibre–matrix interface, resulting in a reduction of mechanical properties (Dlugosz 1965).

Impact strength

The impact strength of fibre-reinforced polymer is governed by the fibre–matrix interfacial bonding, and the properties of both the matrix and the fibres. When the

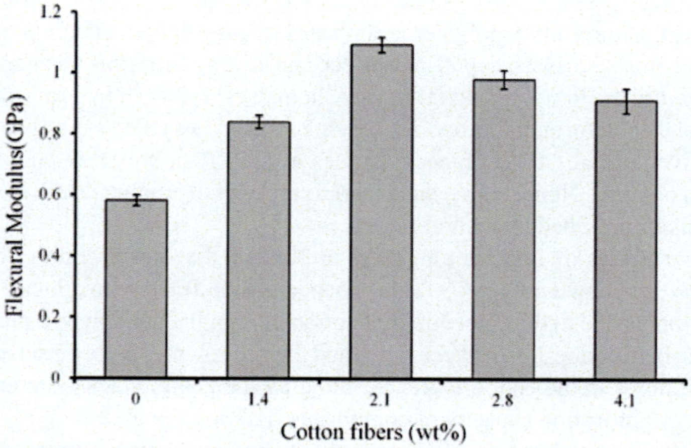

Fig. 4.7 Flexural modulus of geopolymer composites as a function of fibre content

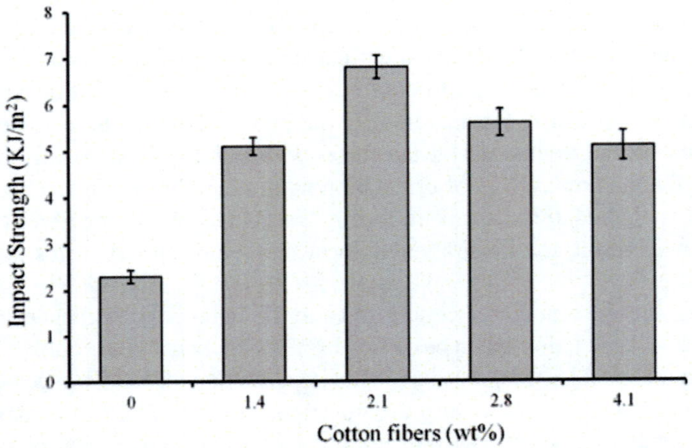

Fig. 4.8 Impact strength of geopolymer composites as a function of fibre content

composites undergo a sudden force, the impact energy is dissipated by the combination of fibre pull-out, fibre fracture and matrix deformation (Wambua et al. 2003). The experimental results presented in Fig. 4.8 indicate that the impact strength of the composites initially increases as cotton fibre content increases to 2.1 wt%, then decreases thereafter. The enhancement in impact strength may be ascribed to good fibre–matrix adhesion, which improves the ability of these composites to absorb impact energy. However, as fibre loading increases, the impact strength significantly decreases. This reduction in impact strength at higher fibre loading is due to their higher porosity and the subsequent formation of voids, which in turn reduced the bond between the fibre and the matrix.

Fracture toughness

In general, natural fibre–polymer composites display crack deflection, de-bonding between fibre and matrix, pull-out effect and a fibre-bridging mechanism, all of which contribute to fracture toughness (Alhuthali et al. 2012). In terms of the matrix alone, plastic deformation provides toughness using an energy dissipation mechanism (Avella et al. 2009; Franco-Marquès et al. 2011), which is hindered by the addition of fibres. Nonetheless, the materials are overall tougher due to the toughness mechanisms provided by natural fibres.

Higher values of fracture toughness are obtained at lower cotton fibre content (2.1 wt%), as shown in Fig. 4.9. This enhancement in fracture toughness at 2.1 wt% cotton fibre is due to the embedding of cotton fibre in the geopolymer matrix, which results in better adhesion between fibres and the geopolymer paste because the spaces between fibres in the cotton fabric are filled by the geopolymer paste and improve the energy absorption capacity of composites (Pakravan et al. 2011).

In contrast, at higher cotton fibre content, there is a reduction in fracture toughness. This is thought to be due to the variation in the amount of geopolymer binder that

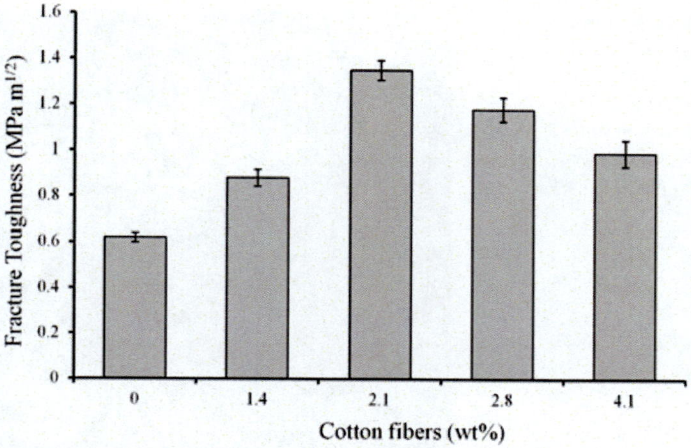

Fig. 4.9 Fracture toughness of geopolymer composites as a function of fibre content

penetrates the openings in the fabric. The penetration of geopolymer binder into the fabric may be maximized when sufficient binder holds the fabric together and gives better adhesion between the fabric and matrix. As the quantity of fabric grows, the amount of binder diminishes, and less is available to penetrate through the fabric openings. As a result, the limited amount of binder penetrating the space of the fabric is insufficient to improve the bonding between the fabric and the matrix. This limitation leads to a reduction in bonding: fibre pull-out occurs readily, and composites exhibit poor toughness results. Therefore, achieving optimal fibre–matrix adhesion is paramount. The results of this study reveal a suitable cotton fibre content that leads to improved fibre–matrix adhesion, with satisfactory fracture toughness and strength.

Microstructure of geopolymer composites

The microstructural analyses of fracture surfaces are shown in Fig. 4.10. It can be observed that the composites with 1.4 and 2.1 wt% cotton fibre show better penetration of the matrix between the fabric openings. This leads to enhancement in the interfacial bonding between the fibre and matrix. However, images in Fig. 4.10c–d clearly indicate that fibre pull-out is quite high, and the bonding between cotton fibre and matrix is very poor. A large gap is evident in the matrix near the cotton fibres. In addition, microcracks can be seen in the fracture surfaces of geopolymer composites with 4.1 wt% cotton fibre, which confirms that fibre–matrix de-bonding has occurred, and thus strength has been reduced. This is clear evidence that the fibre matrix interfacial adhesion is better for CFGs with 1.4 and 2.1 wt% than for those with 2.8 and 4.1 wt%. The data presented above for the mechanical properties of geopolymer composites is also supported by the SEM observations.

Fig. 4.10 SEM images of the fracture surface for geopolymer composites reinforced with varying content of cotton fibres: **a** 1.4 wt%, **b** 2.1 wt%, **c** 2.8 wt%, and **d** 4.1 wt%

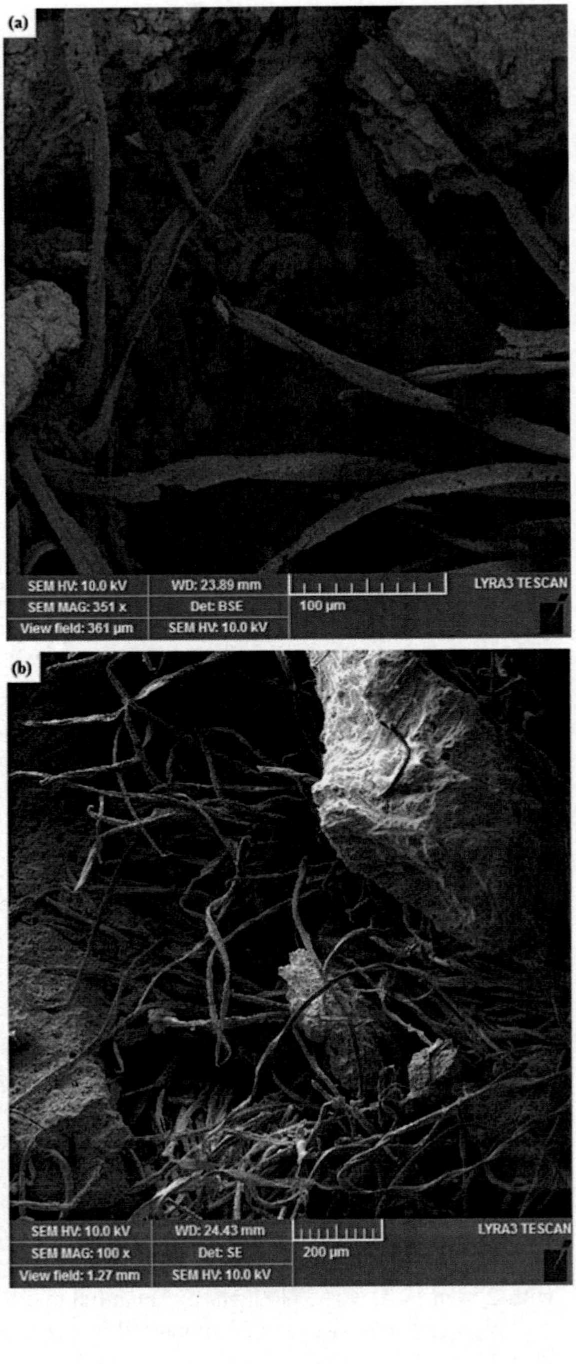

Fig. 4.10 (continued)

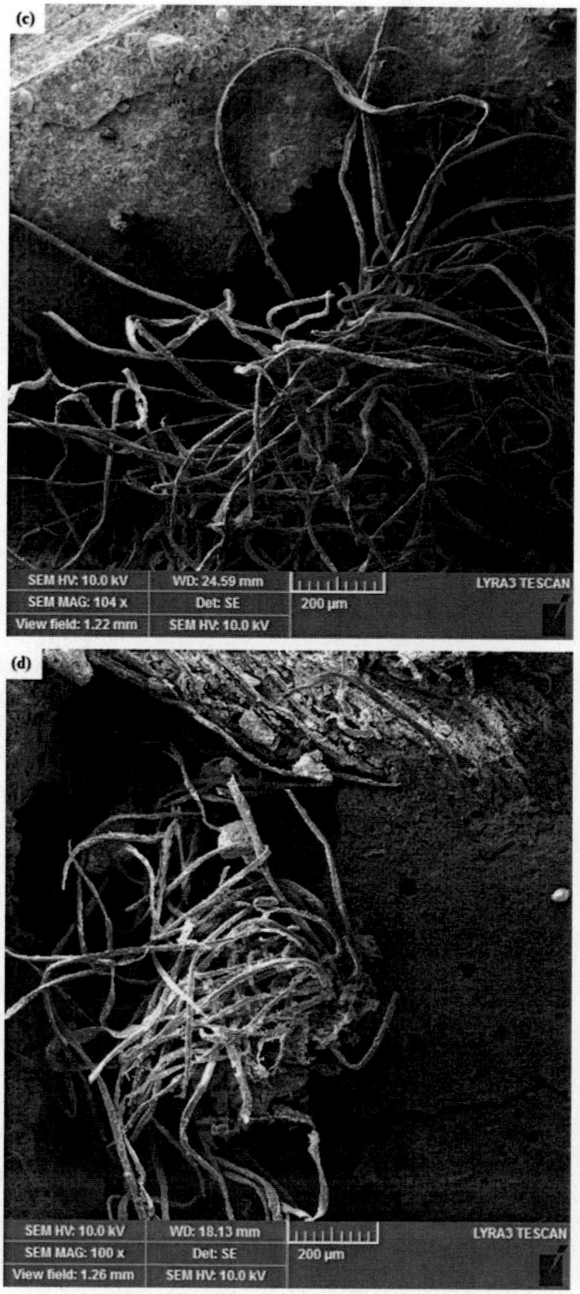

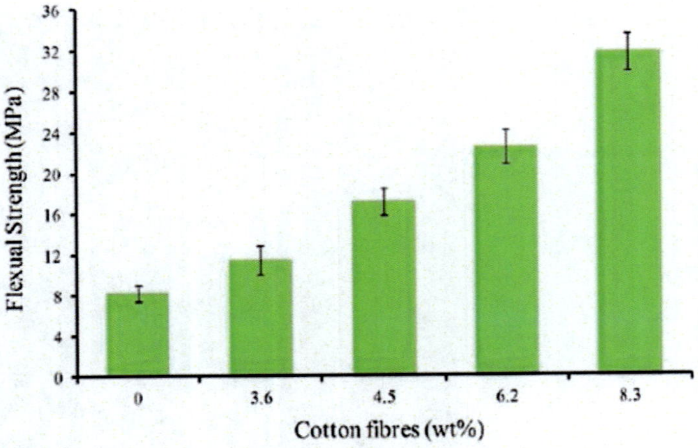

Fig. 4.11 Flexural strength of geopolymer composites as a function of fibre content

(ii) ***Forced-impregnation hand lay-up technique***

Flexural strength and modulus

Flexural tests are often used to characterise the mechanical properties of layered materials since they provide a simple means of determining the bending response. This provides useful information on the performance of layered fabric-based composites (Abanilla et al. 2006). The test results show that the flexural strength of cotton fabric reinforced composite increases as the wt% of cotton fibres increases (see Fig. 4.11). The composite containing 8.3 wt% woven cotton fibres exhibited the highest flexural strength among all composites. The flexural strength of the composites increased from 8.2 to 31.7 MPa compared to pure geopolymer. This indicates that increasing the number of woven cotton fibres leads to considerable improvement in flexural strength in the composite. This finding can be justified from the fact that the flexural strength is controlled by the number of reinforcement layers. The lower weight of cotton fabrics allows multiple *layers* of fabric in the composite, *to resist the shear failure and contribute in sustaining the applied load to the composites. This permits greater stress transfer between the matrix and the cotton fibres, resulting in improved flexural strength* (Sim et al. 2005).

In previous studies (Alomayri and Low 2013; Alomayri et al. 2013a, b), the authors studied the flexural behaviour of short cotton fibre reinforced geopolymer composites and observed minimum improvement in the flexural strength over that of the unreinforced specimens due to poor dispersion of short cotton fibres in the matrix. In fact, agglomerations of cotton fibres were noticed which degraded the interfacial adhesion between the fibre and the matrix as shown in Fig. 4.15. In present study, the utilisation of continuous fibres as reinforcement for geopolymer composites has shown better mechanical properties than short cotton fibres, owing to their ability to effectively bridge the cracks due to their alignment in the direction of tension which

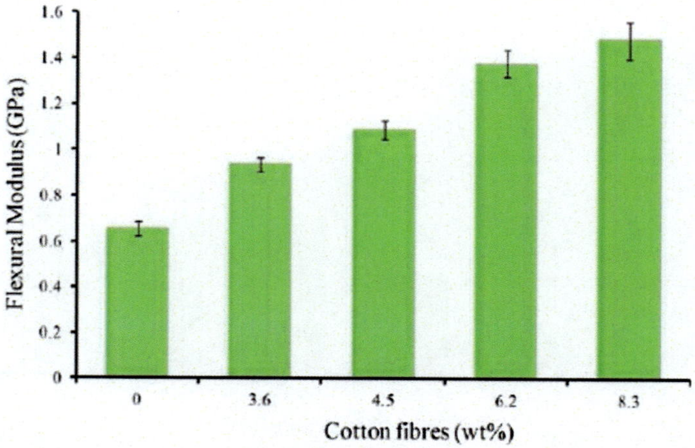

Fig. 4.12 Flexural modulus of geopolymer composites as a function of fibre content

resulted in greater stress transfer at the interface of the composites. The improved performance of cotton fabric-geopolymer composites can be explained by observing the SEM microstructure images as shown in Fig. 4.7f. This shows good penetration of geopolymer paste into the filament of the cotton bundle making up the fabric, thus providing improved bonding between the fabric and the geopolymer matrix and leading to an improvement in flexural strength.

The flexural modulus of geopolymer composites is shown in Fig. 4.12 and indicates similar trends to flexural strength values. The addition of woven cotton fibres in the geopolymer matrix increases the flexural modulus over plain geopolymer matrix. The flexural modulus is a measure of resistance to deformation of the composite in bending. It was observed that none of the specimens are completely broken at peak load. This could be due to the crack bridging by long continuous fibres, which makes their flexural modulus higher than un-reinforced geopolymer.

Such fibres can withstand a higher load and can undergo multiple cracks throughout the loading process, thus preventing brittle failure of the specimens. Similar results have been reported by Low et al. (2007) when testing the mechanical properties of cellulose fibre-reinforced epoxy laminates using the three-point bending tests. They reported an increase in both flexural strength and modulus as the fibre contents increase. The increase in woven cotton fibre content was exceptionally useful in terms of improving the flexural strength and modulus of this inorganic polymer matrix.

Impact strength

Impact strength is an essential dynamic property of engineering material that gives an indication of its resistance against sudden impact. The impact strength of fibre reinforced polymer is governed by the matrix-fibre interfacial bonding, and properties of matrix and fibres. When the composites undergo a sudden force, the impact energy is

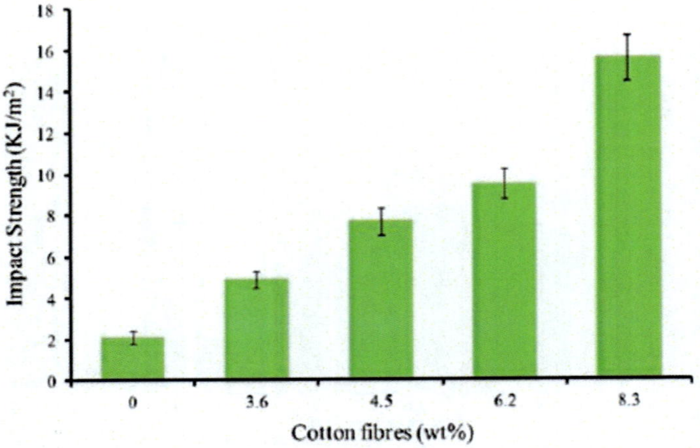

Fig. 4.13 Impact strength of geopolymer composites as a function of fibre content

dissipated by the combination of fibre pull outs, fibre fracture and matrix deformation (Mishra et al. 2003; Alamri and Low 2012a, b, c). The experimental results presented in Fig. 4.13 indicate that the impact strength of the composites increases as cotton fibre content increases. The impact strength of the neat geopolymer increased from 2.1 to 15.6 kJ/m^2 after the addition of 8.3 wt% woven cotton fabric to the geopolymer composite. This significant enhancement in impact strength could be attributed to the use of applied load on the top surface of the geopolymer composites during sample preparation, which expelled the trapped air from the sample and forced the geopolymer paste into the voids and pore spaces. As a result, the bonding between the fabrics and the matrix is enhanced, and results in increased impact strength.

This improvement in impact strength may also be attributed to the fibrillation of cotton layers that creates branches in the fibre (see Fig. 4.16g), resulting in the formation of micro fibrils which increase the fibre specific surface area (Nakagaito and Yano 2004; Zhang et al. 2005). This leads to an enhanced fibre-matrix interaction, and a strong bonding between the micro fibrils and the geopolymer matrix. Therefore, better stress transfer from the matrix to the micro fibrils results in increased fibre-matrix bonding. Similar remarkable improvements in impact strength were reported by Graupner (2008) where the addition of cotton fibre increased the impact strength of pure poly (lactic acid) (PLA) matrix. He concluded that the increase was due to greater elongation of cotton fibres at break. Natural Fibres containing cellulose generally have high elongation at break values. Cotton fibres have a cellulose content of about 88–96%. Generally, elongation at break and impact strength are directly correlated. The high elongation at break of cotton fibres increased the elongation at break in the composites, leading to higher impact strength (Figs. 4.14 and 4.15).

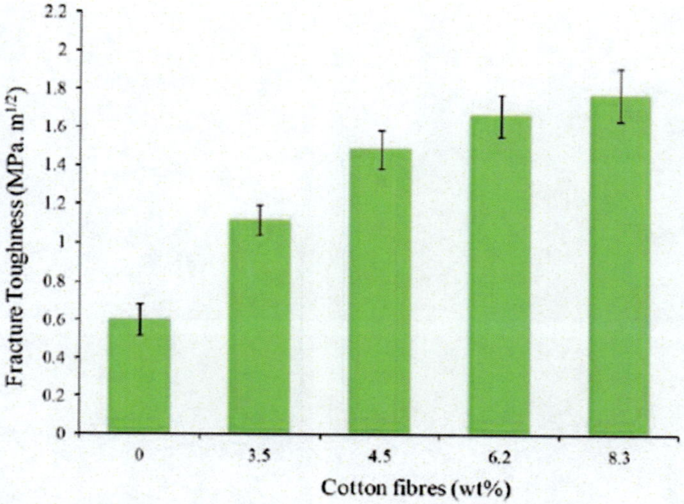

Fig. 4.14 Fracture toughness of geopolymer composites as a function of fibre content

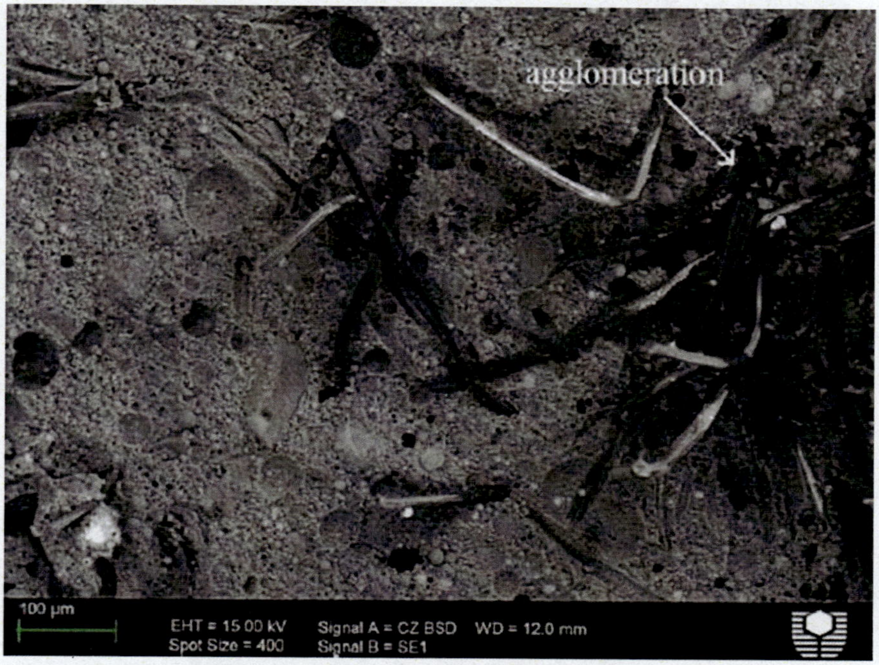

Fig. 4.15 SEM showing agglomeration of short fibres in cotton fibre/geopolymer composites loaded with 1.0 wt%

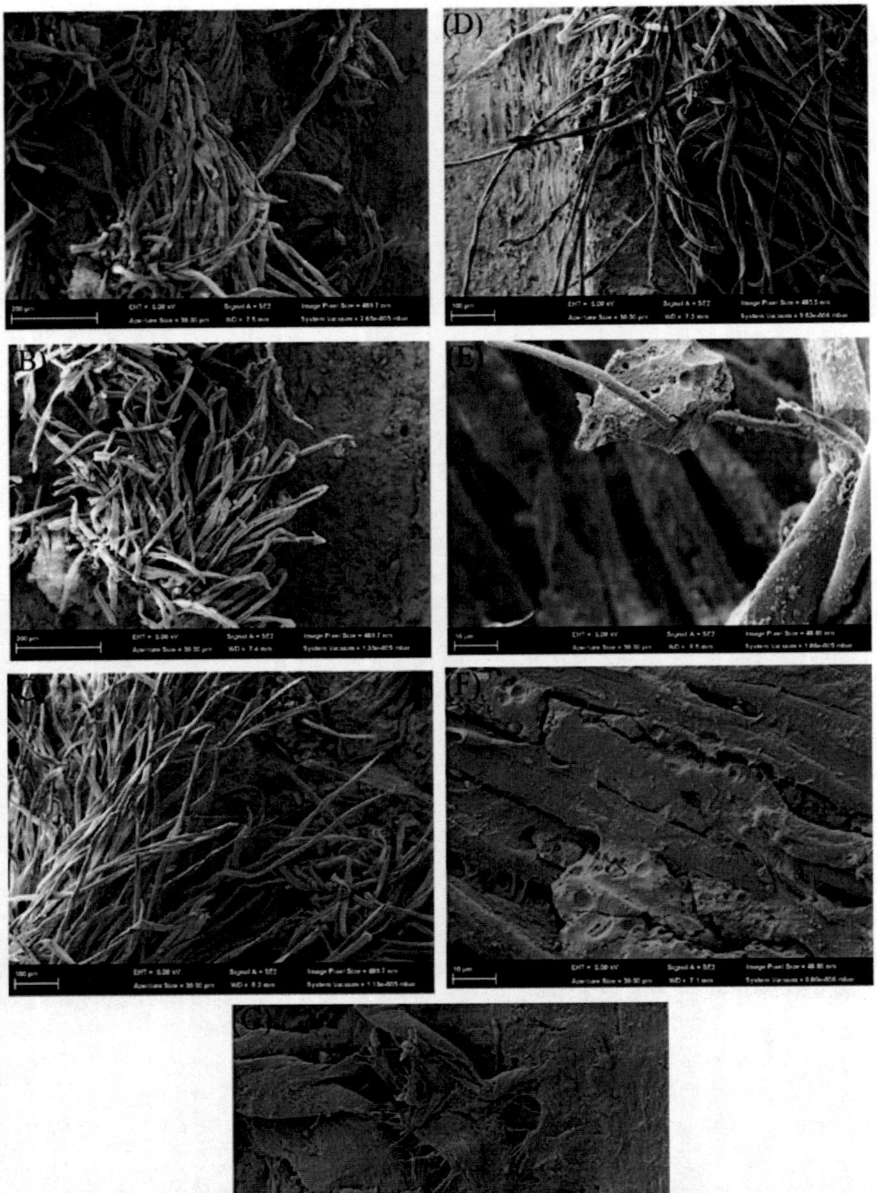

Fig. 4.16 SEM images of the fracture surface for geopolymer composites reinforced with varying content of cotton fibres **a** 3.6, **b** 4.5, **c** 6.2 and **d–e** 8.3 wt%. The micrographs of **f** penetration of the geopolymer matrix into cotton fabrics and **g** micro fibrillation of cotton fibres

Fracture toughness

Generally, crack deflection, debonding and bridging of fibres slows down the crack propagation in fibre reinforced composites and increase the fracture energy (Reis 2006; Silva et al. 2009, 2010; Tolêdo Filho et al. 2000, 2003; Joseph et al. 1999). Figure 4.5 shows the influence of cotton fabric content on the fracture toughness of the composites, the composite containing higher cotton fibre content exhibits higher fracture toughness. The greatest improvement in fracture toughness (from about 0.6 MPa $m^{1/2}$ in the unreinforced matrix to about 1.8 MPa $m^{1/2}$) was obtained with 8.3 wt% cotton fibre reinforcement. This extraordinary enhancement is due to the unique properties of woven cotton fibre to resist fracture resulted in increased energy dissipation from crack-deflection at fibre–matrix interface, fibre-debonding, fibre-bridging, fibre pull-out and fibre-fracture. The high values of fracture toughness obtained in geopolymer composites with woven cotton fibres were due to better interface interaction between fibre and matrix as shown in Fig. 4.16. The improved interfacial adhesion enabled higher stress transfer between the fibres and matrix and reduced the chance of fibre de-bonding. Accordingly, the load required to break the sample increases when the content of cotton fibre is increased. Therefore, the fracture toughness of geopolymer composites increases with increasing wt% of cotton fibres.

Figure 4.16e demonstrates selected scanning electron micrographs of fracture surface of geopolymer composite and explains the fracture toughness behaviour. It can be seen from Fig. 4.16e that small pieces of geopolymer paste were attached to the fibre surface of cotton fibre/geopolymer composites. Hence, retention of the matrix on the fibre surfaces shows the good adhesion between cotton fibres and geopolymer matrix. Additionally, it was observed that geopolymer composites with woven cotton fibres did not completely break into two pieces due to close spacing of woven cotton fabric which lead to fibres bridging the cracks and enhancing the crack propagation resistance. The tortuous pathway of the crack propagation indicates that high energy is absorbed by the cotton fibre layers (see Fig. 4.17).

Microstructure characteristics

The fracture surfaces of the woven cotton fibre reinforced geopolymer composites have been studied under SEM and are shown in Fig. 4.16. Generally, fibre pullout, fibre-debonding, fibre breakage and matrix fracture are observed after the fracture test of all composites. In fact, such toughening mechanisms increased the fracture properties of samples reinforced with woven cotton fibres. The effect of fibre content on the fracture surface is clearly seen in images of Fig. 4.16a, b. Composites filled with lower fibre content (3.6 and 4.5) wt% show an increase in matrix-rich regions compared to composites filled with higher fibre content. An increase in matrix rich regions means that the matrix is not reinforced by enough fibres. Therefore, there are insufficient fibres to transfer the load from the matrix (Simonova et al. 2018; Joseph et al. 1999). Due to this reason, the geopolymer composites with low fibre content exhibited low fracture toughness and mechanical properties. The fracture surfaces of the geopolymer composites with higher fibre content are shown in Fig. 4.16c, d. There are higher fibre-rich regions of composites filled with 6.2 and 8.3 wt% cotton

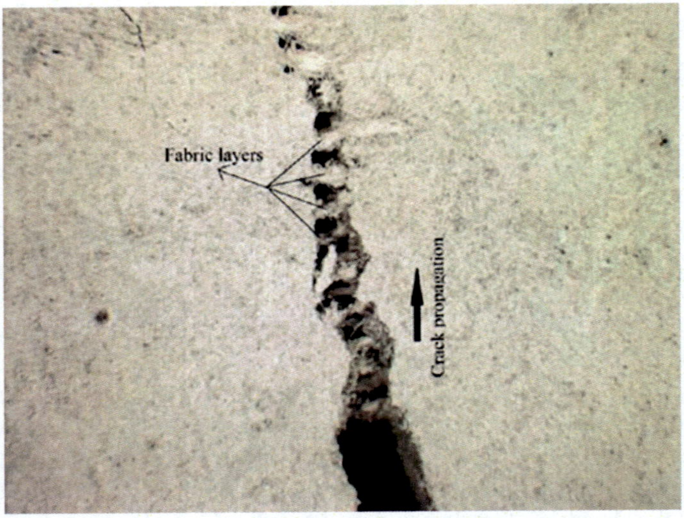

Fig. 4.17 Optical image of crack propagation in cotton fabric/geopolymer composites loaded with 8.3 wt% fibre

fibres. An increase in fibre-rich regions means greater stress-transfer from the matrix to the cotton fibres thereby resulting in the improvement of mechanical properties. Therefore, these observations indicate that the woven cotton fabrics can be used as potential material to reinforce geopolymer composites due to their good mechanical properties.

These results clearly show that the presence of cotton fabric layers in the geopolymer composites significantly increases the flexural strength, flexural modulus, impact strength and fracture toughness when compared to neat geopolymer. This remarkable enhancement is due to the unique properties of cotton fibres in withstanding greater bending and fracture forces than the more brittle geopolymer. SEM micrographs show several toughness mechanisms which include crack bridging, fibre pullout and fibre fracture, and matrix fracture. These toughening mechanisms are the major factors contributing to the enhanced mechanical properties of cotton fabric-reinforced geopolymer composites. Cotton fibres appear to be uniquely suited to reinforce geopolymer composites since they can be easily processed using conventional manufacturing techniques to yield a product with good mechanical properties at low cost. They can be classified as desirable performing composites that offer benefits to engineers, particularly in less developed countries or countries that need low-cost construction materials. Possible applications for cotton fibre-reinforced geopolymer composites include slabs or shingles for siding, certain types of roofing, and some interior uses in the building structure. They may also be used for other applications such as pipes and cooling towers.

4.2.2 *Effect of Fabric Orientation*

Table 4.3 shows the compositions of the fabricated samples whereby each sample contained different layers of cotton fabric. For each sample, the final layer was geopolymer paste. During the testing of mechanical properties, the samples were tested in two directions as the fabrics were aligned either horizontal or vertical to the applied load (see Fig. 4.18).

Flexural strength

The effect of cotton fibre content and fibre orientation on flexural strength is shown in Fig. 4.19. It is observed that the incorporation of cotton fabric layers has significantly increased the flexural strength of the composites. Thus, there is a significant difference between the flexural strength of cotton fabric reinforced geopolymer composites and the control specimens. This is observed in both horizontal and vertical directions of the cotton fabrics, indicating the advantage of using cotton fibres to reinforce geopolymer composites as previously discussed in earlier publications (Alomayri et al. 2013a, b). Figure 4.19 also shows the effect of fabric direction on the flexural strength of the composites. The results showed that the composites with horizontal fabric layers (i.e., load normal to fabric layers) have higher flexural strength than

Table 4.3 Formulations of samples

Sample	Fabric layers	Fibre content (wt%)
Composite 0	0	0
Composites 1	10	4.5
Composites 2	20	6.2
Composites 3	40	8.3

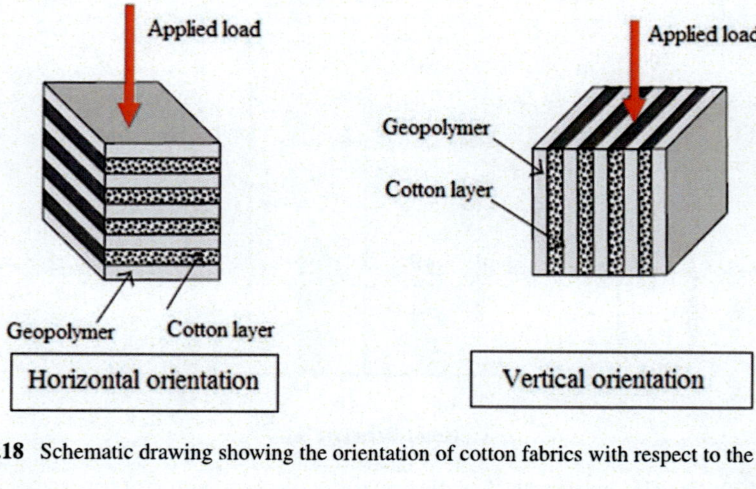

Fig. 4.18 Schematic drawing showing the orientation of cotton fabrics with respect to the applied load

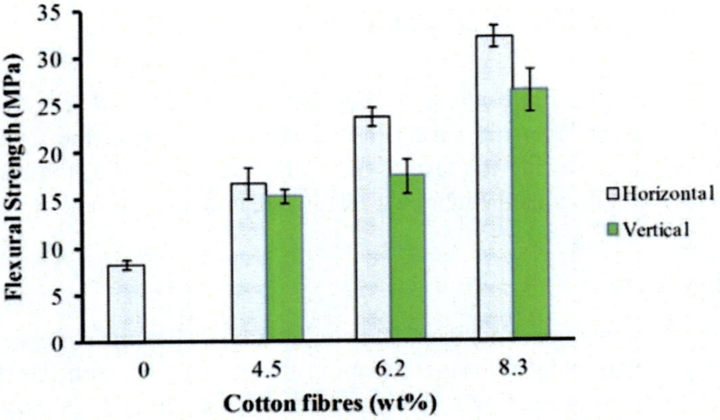

Fig. 4.19 Flexural strength of geopolymer composites as a function of fibre content

those with vertical fabric layers (i.e., load parallel to fabric layers). The higher flexural strength in composites with horizontally laid cotton fabric can be attributed to better uniformity in load distribution among the consecutive layers of cotton fabric.

Compressive strength

The compressive strength of geopolymer composites containing cotton fabrics laid in both horizontal and vertical directions are presented in Fig. 4.20. The results showed that the compressive strengths are affected significantly by the fabric direction. The compressive strength of composites is higher in the case of horizontally oriented fabric compared to that laid in vertical direction. This can be due to the ability of horizontally laid cotton fabric to directly absorb and distribute the load uniformly throughout the cross-section (Sankar et al. 2017; Alomayri and Low 2013). In addition, this significant enhancement of compressive strength in the horizontal direction

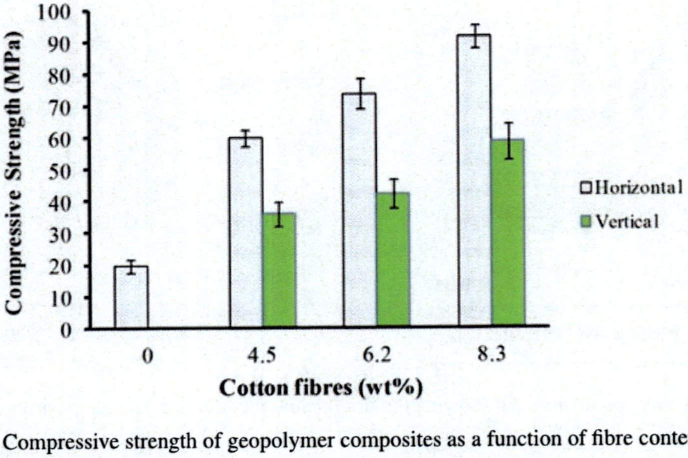

Fig. 4.20 Compressive strength of geopolymer composites as a function of fibre content

is because the interface between the fabric and the matrix is not exposed to any shear loading which in turn reduces the possibility of fabric detachments or delamination from the matrix at high loads. However, the scenario is different when fabric is oriented vertically to the compressive load where delamination between the cotton fabric and matrix can happen, which will result in inefficient stress transfer between the fabric and matrix.

Moreover, the results in Fig. 4.20 show that the compressive strength of the composites containing cotton fabric increases with increase in fabric layers (i.e. the fibre contents) oriented in both directions. The increase in compressive strength with fibre loading may be due to the ability of the cotton fibres to absorb the stress transferred from the matrix. Therefore, the results of compression test in this study revealed that addition of cotton fabric enhances the compressive strength of fly ash-based geopolymer composites.

Hardness

The hardness values of cotton fabric reinforced geopolymer composites are shown in Fig. 4.21. The results show that the hardness of composites increases with increase in the fibre loading. This is true for both horizontally and vertically laid fabrics. It can also be seen that the hardness of composites with horizontally laid cotton fabric exhibited slightly higher hardness than those containing vertically oriented fabric. This is due to the uniform distribution of the load on cotton fibres which decreased the penetration of the test ball to the surface of the composite and consequently increased the hardness of composite (Alomayri and Low 2013).

Fracture toughness

In general, processes such as crack deflection, debonding and bridging of fibres will slow down the crack propagation in fibre reinforced composites and increase the fracture energy (Zhang and Li 2004; Reis 2006; Silva et al. 2009, 2010). Fracture toughness of geopolymer composites containing cotton fabric in horizontal and

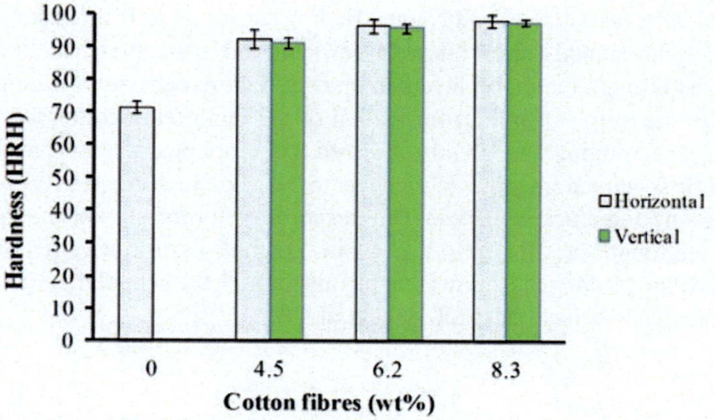

Fig. 4.21 Hardness of geopolymer composites as a function of fibre content

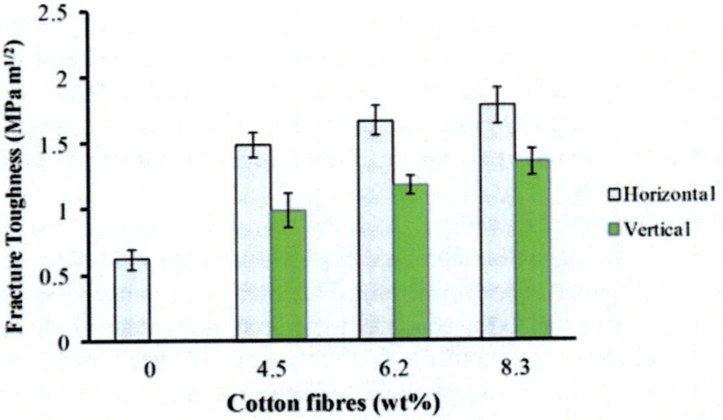

Fig. 4.22 Fracture toughness of geopolymer composites as a function of fibre content

vertical directions are shown in Fig. 4.22. The fracture toughness increases as the content of cotton fibre increases. This significant enhancement of facture toughness is due to fibre pull-out, fibre fracture and fibre-bridging, as clearly shown in the SEM images of Fig. 4.24a–c.

Figure 4.23a, b shows the flexural stress–strain curves of geopolymer composites containing cotton fabric in horizontal and vertical directions. The composites containing horizontally laid cotton fabrics exhibited non-catastrophic fracture behaviour. All horizontally reinforced cotton fabric reinforced composites exhibited strain hardening behaviour as can be seen in Fig. 4.22a. The flexural strength and strain at peak load increase with increase in cotton fabric layers. On the other hand, the composites containing vertically aligned cotton fabrics showed some nonlinearity at very low strain, followed by strain softening after the peak load. This suggests the feasibility of using horizontal layers of cotton fabrics to mitigate the brittle failure of geopolymers. Moreover, the areas under the curves give an indication that the composite containing horizontal cotton fabric layers achieved higher fracture toughness than samples with vertically aligned cotton fabric layers. In the case of composite containing horizontal cotton fabric layers, the fabrics are stretched, and the crack developed through the fabric layers in graceful failure behaviour, leading to a high degree of ductility. Also, the contribution of the reinforcing cotton fabric to crack arresting and bridging were clearly observed. The crack propagates through the thickness of the specimen from one fabric layer to the next. Such crack arresting, bridging mechanisms and crack deflection were responsible for the significant enhancement in fracture toughness as seen in Fig. 4.24d. On the contrary, at vertical loading the crack developed along the fabric layer mainly through the geopolymer matrix leading to a more brittle behaviour (Mobasher et al. 2006).

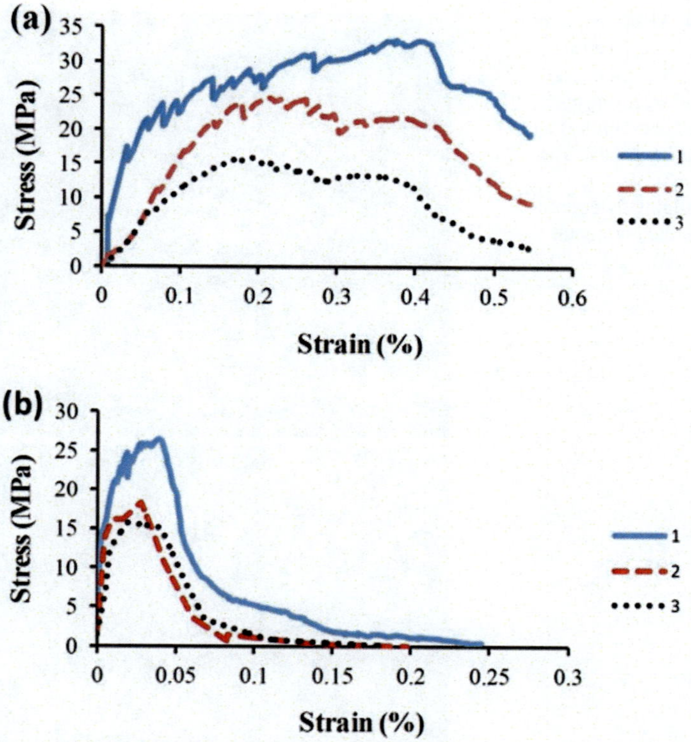

Fig. 4.23 Typical stress–strain curves of geopolymer composites with cotton fabrics tested: **a** horizontally and **b** vertically to the applied load [Legend: 1 = 8.3 wt%, 2 = 6.2 wt%, 3 = 4.5 wt%]

Similarly, this enhancement of fracture toughness in the horizontal direction has been reported by other researchers when dealing with natural fibre-based composites. Low et al. (2009) reported that cellulose fibre-reinforced epoxy composites in the horizontal direction to the applied load achieved higher fracture toughness when compared to samples with fibre sheets in the vertical direction to the applied load. Thus, they concluded that the higher fracture toughness in the horizontal could be attributed to the pronounced display of interfacial crack-deflection, leading to a very tortuous crack path. Hence, the composite sample failed in a more graceful manner with discontinuous or multiple "stick–slip" fracture. In contrast, the composite samples in the vertical direction showed continuous crack growth. Therefore, the phenomenon of multiple "stick–slip" fracture is attributed to the repeating occurrence of crack initiation, arrests and de-bonding at the cotton fabric/geopolymer interfaces.

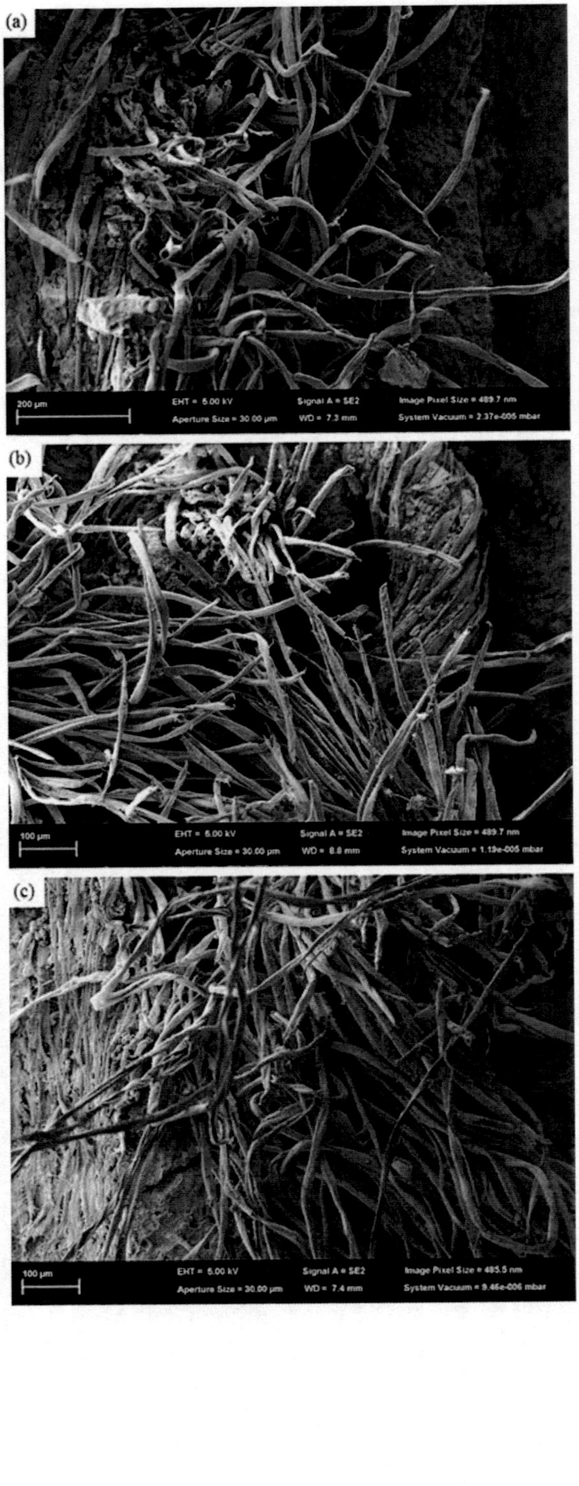

Fig. 4.24 SEM images of the fracture surface for geopolymer composites reinforced with varying content of cotton fibres **a** 4.5, **b** 6.2 and **c** 8.3 wt%. The micrographs of crack-bridging by cotton fabrics in the horizontal direction (**d**)

Fig. 4.24 (continued)

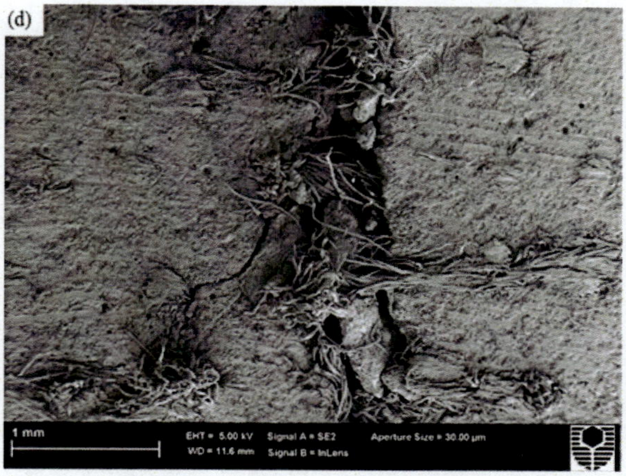

4.2.3 *Effect of Nanoclay*

Mechanical properties

The flexural tests are often used to characterize the mechanical properties of composites as they provide a simple means of determining the bending response. This provides useful information on the performance of the composites. The effect of nanoclay addition on the flexural strength of geopolymer nanocomposites is presented in Fig. 4.25. Experimental results indicate that the flexural strength of samples initially

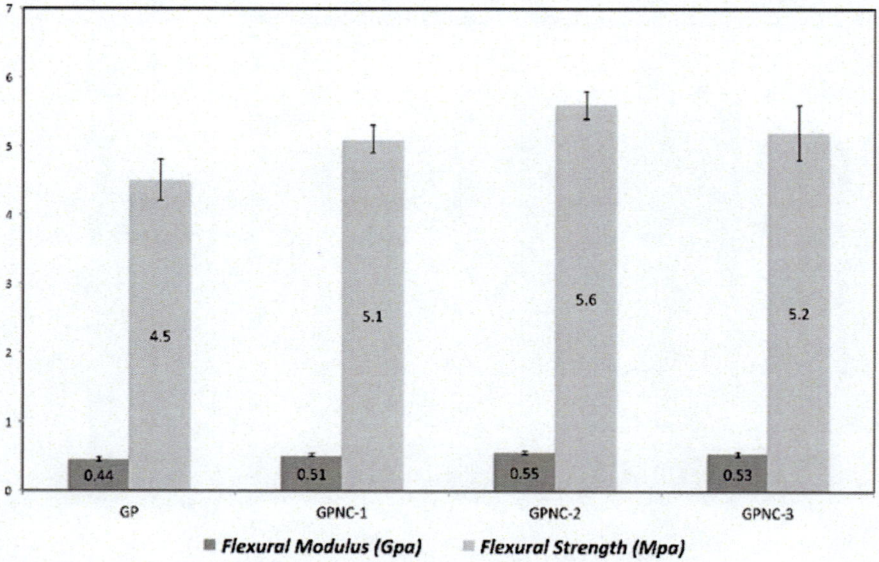

Fig. 4.25 Flexural strength and flexural modulus of samples GP, GPNC-1, GPNC-2 and GPNC-3

increases with increasing nano-clay content of up to 2.0 wt% but decreases at higher contents. The flexural strength of the nanocomposites is improved from 4.5 MPa in the control to about 5.6 MPa with 2.0 wt% nano-clay. This improvement can be attributed to the good dispersion of the nanoparticles throughout the matrix, leading to less porosity and denser geopolymer matrix. This result is comparable with that of nano-clay-cement composite reported by Hakamy et al. (2013a, b) who found that the optimum addition of nano-clay to cement mix is about 1.0 wt%, which increases the flexural strength of the nano-composites by 31% over the control sample. Both studies imply that increasing the content of nano-clay led to some improvement in flexural strength of the composite. This result can also be ascribed to the nano-particles effect, which improved the matrix through geopolymeric reaction and increased the amorphous content, producing higher content of geopolymer products.

The flexural modulus is a measure of resistance to deformation of the composite in bending. The flexural modulus of control pastes and geopolymer nanocomposites is shown in Fig. 4.25 and indicates that the optimum addition of nano-clay is 2.0 wt% to the geopolymer matrix, and it improved the flexural modulus over a pure geopolymer matrix by 25%.

The compressive strength results of geopolymer and geopolymer nanocomposites are shown in Fig. 4.26 and indicate similar trends to flexural strength and modulus values. Compressive strength is inversely proportional to porosity: specimens with less porosity displayed higher compressive strength. The compressive strength of the neat geopolymer paste is improved from 37.2 to 45.9 MPa after the addition of 2.0 wt% nano-clay, but this trend is reversed, reducing the strength to 40.2 MPa

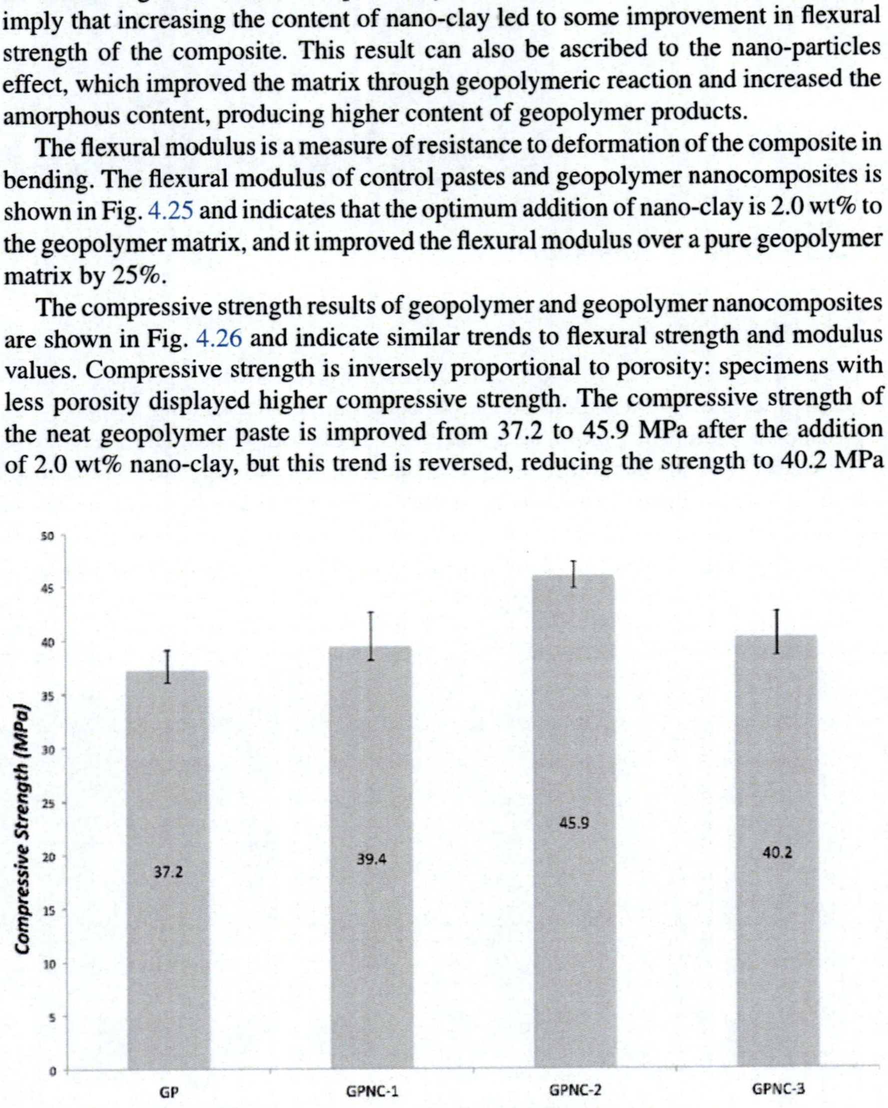

Fig. 4.26 Compressive strength of samples GP, GPNC-1, GPNC-2 and GPNC-3

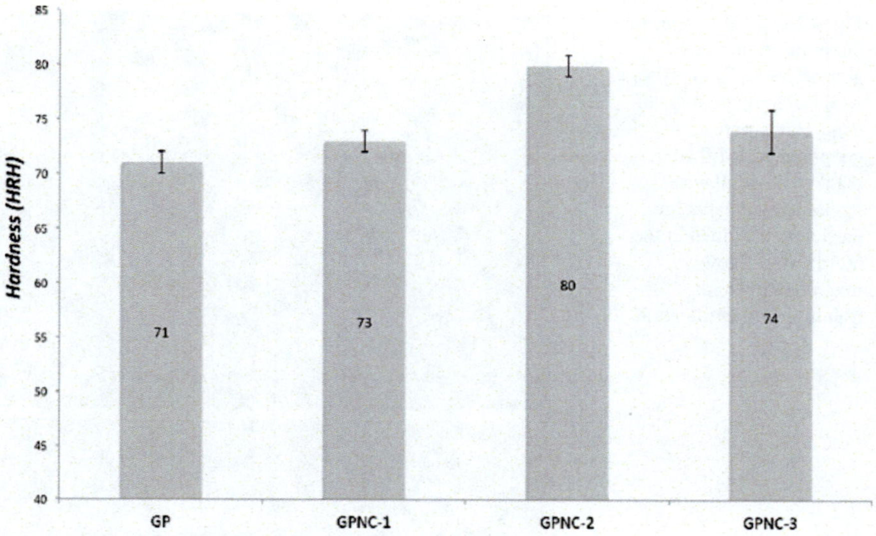

Fig. 4.27 Hardness of samples GP, GPNC-1, GPNC-2 and GPNC-3

with the addition of 3.0 wt% nano-clay. In a similar study, Phoo-ngernkham et al. (2014) reported that the addition of 1.0–2.0 wt% nano-alumina and amorphous nano-silica into geopolymer matrix enhanced the geopolymeric reaction and increased the geopolymer gel, which increased the density and consequently improved the compressive strength of geopolymer matrix. Both studies showed that increasing the compressive strength of geopolymer pastes is corresponded to the reduction in porosity.

The hardness values of the control sample and geopolymer nanocomposites are presented in Fig. 4.27. The results show that there was no significant improvement observed between all samples. However, the geopolymer nanocomposite with 2.0 wt% of nano-clay showed slightly higher hardness than other samples. This enhancement could be attributed to the high density of the geopolymer nano-composite paste, which decreased the penetration of the test ball on the surface of the nano-composite matrix and consequently improved the hardness.

SEM observation

SEM images in Fig. 4.28a–d show the fracture surfaces of nanocomposite containing 0, 1.0, 2.0, and 3.0 wt% nano-clay. The pure geopolymer has a less dense matrix with a higher number of non-reacted and partially reacted fly-ash particles embedded in the matrix (Fig. 4.28a). For the 1–3 wt% nano-clay (Fig. 4.28b–d) less numbers of fly-ash particles were observed, and the matrix seemed denser than that of the control paste. Images in Fig. 4.28e–f display an observation of the geopolymer matrix that loaded with 3.0 wt% nanoclay at low magnification. Nanoclay particles are poorly dispersed and agglomerated due to the high content of nanoclay. Figures 4.28g–h show agglomerations of nanoclay platelets at higher magnification.

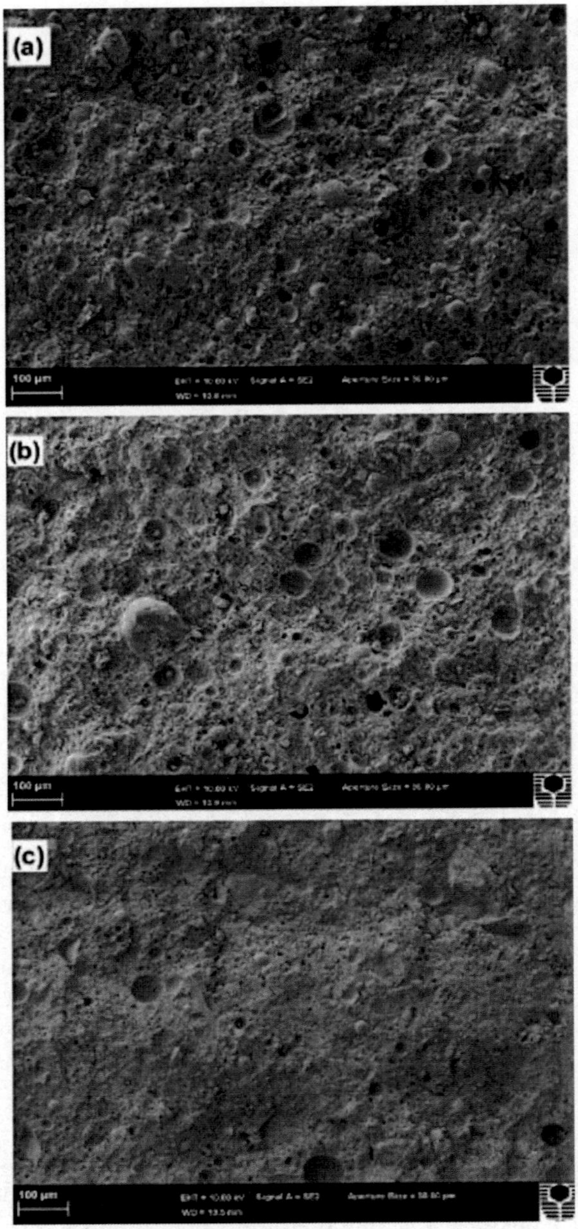

Fig. 4.28 SEM images of the fracture surface of geopolymer nanocomposites with different loadings of nano-clay **a** pure geopolymer, **b** 1.0 wt%, **c** 2.0 wt%, **d** 3.0 wt%, agglomerated nanoclay particles embedded in the matrix at: **e–f** low magnification and **g–h** higher magnification

Fig. 4.28 (continued)

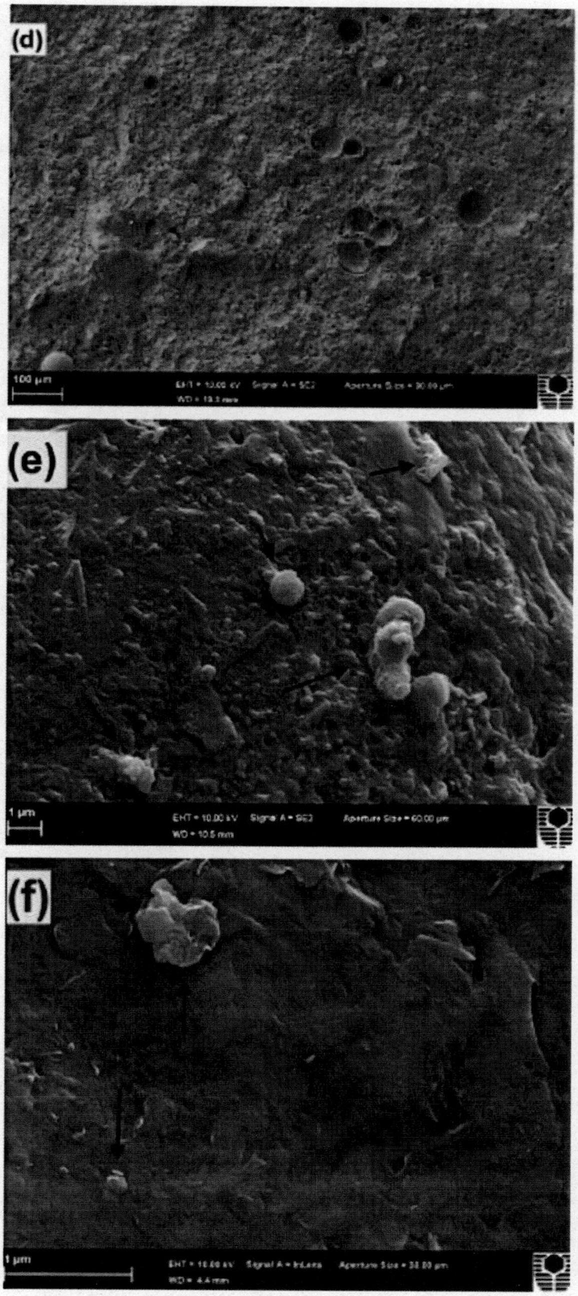

Fig. 4.28 (continued)

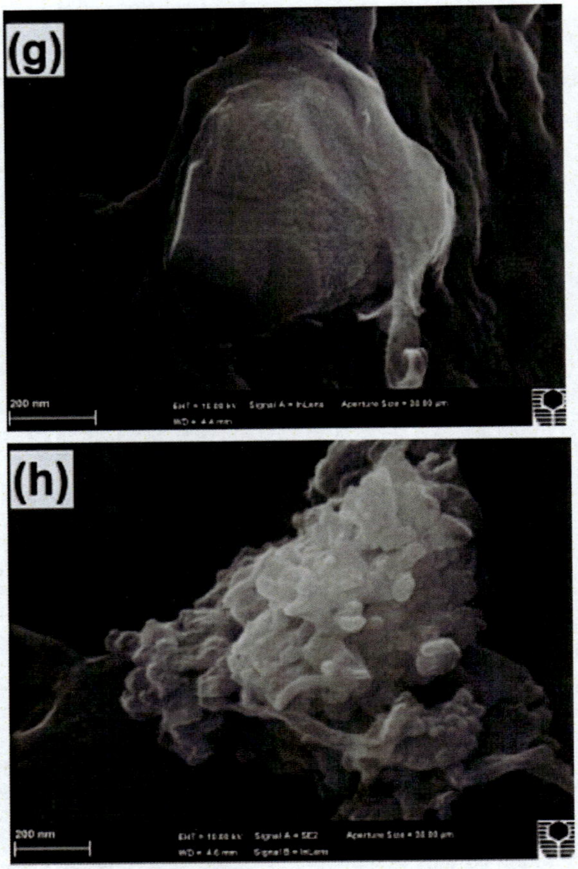

4.2.4 *Effect of Ordinary Portland Cement*

Flexural strength

The flexural strength of the geopolymer matrix and CF-reinforced geopolymer composites containing different wt% of OPC is shown in Fig. 4.29. In general, the incorporation of OPC into the geopolymer matrix led to an enhancement of flexural strength, as shown in the figure. The addition of 10 wt% OPC resulted in the highest flexural strength of all the geopolymer samples. The flexural strength of the geopolymer matrix containing 10 wt% OPC was increased from 9.3 to 10.7 MPa compared to pure geopolymer. This improvement in mechanical properties can be attributed to two mechanisms. First is the filling effect, where the OPC fills the voids or pores in the geopolymer paste, making the microstructure of the matrix denser than that of the geopolymer paste without OPC. The second mechanism is the pozzolanic reaction, in which the OPC with fly ash produces C–S–H gel in the geopolymeric system, which may work as a micro-aggregate and reduce porosity, thereby enhancing

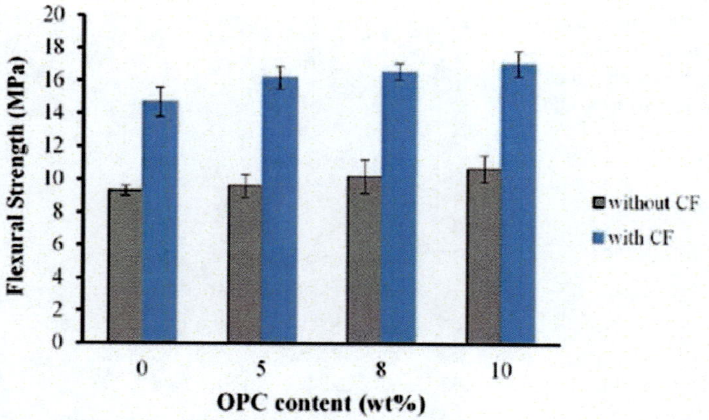

Fig. 4.29 Flexural strength as a function of OPC content for geopolymers with and without CF

the mechanical properties (Yip and van Deventer 2003; Yip et al. 2005; Dombrowski et al. 2007; Boonserma et al. 2012). X-ray diffraction analysis reveals evidence of a calcium silicate hydrate (C–S–H) phase that may also contribute to increased flexural strength in the samples with added OPC. The SEM–EDS provides evidence of the formation of C–S–H within the geopolymeric gels as shown previously in Fig. 3.10.

The higher flexural strength of samples with added OPC compared to those without is a result of the increased reactive amorphous phase of the mixture, resulting from the blending of fly ash with OPC. Figure 4.30 shows micrographs of geopolymers with no added OPC and 10 wt% OPC, cured at ambient temperature. The micrographs show that the addition of OPC causes a fine, more homogeneous microstructure (see images in Fig. 3.10b, c); while for geopolymer with no added OPC there is evidence of partially reacted fly ash spheres (see Fig. 3.10a). Increasing the OPC content created a more compact and finer microstructure, indicating that OPC is acting as a seeding or precipitating element. The probable reaction product is calcium silicate hydrate (C–S–H), which has been detected by XRD. Previous studies (van Jaarsveld et al. 1999; Phair and van Deventer 2001) have found that C–S–H gel has a positive impact on the mechanical strength of geopolymeric binders. The current findings agree with these studies. It is confirmed that the addition of OPC increases the mechanical properties of geopolymer matrix due to the formation of C–S–H gel in the geopolymer system.

The flexural strengths of CF-reinforced geopolymer composites are also shown in Fig. 4.29. The presence of CF significantly improved the flexural strength for all samples. Additionally, the flexural strength of neat geopolymer increased from 9.3 to 14.7 MPa after the addition of CF. This enhancement in flexural properties is clearly the contribution of the cotton fabrics in the composites. The inclusion of OPC to the CF-reinforced geopolymer composite also increased flexural strength; the flexural strengths of composites containing 5, 8 and 10 wt% OPC were 16.2, 16.6 and 17.1 MPa, respectively. The increase in flexural strength of geopolymer composites after the addition of OPC can be attributed to the enhancement in the

Fig. 4.30 SEM micrographs showing the microstructures of geopolymer with **a** 0 wt% OPC, and **b**, **c** 10 wt% OPC

Fig. 4.30 (continued)

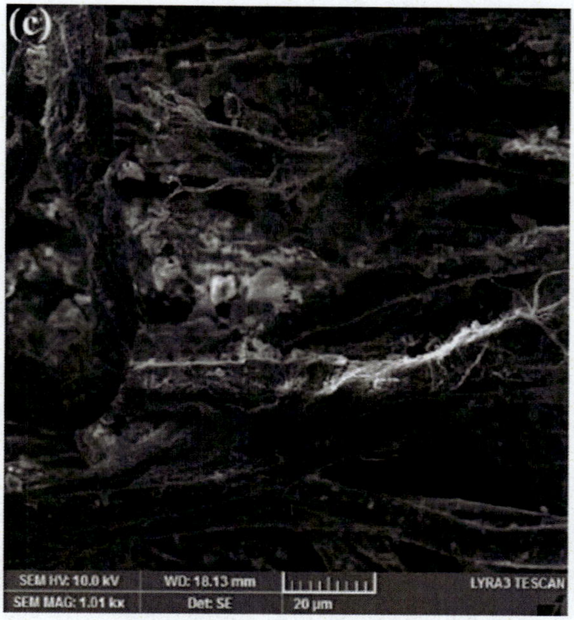

interfacial adhesion between the fibre and the matrix. Good interfacial bonding results in enhanced composites strength properties, since stress can be effectively transferred between the fibre and the matrix. This indicates that the addition of OPC is useful for improving the flexural strength of inorganic polymer composites.

Impact strength

The impact strength of geopolymer matrix containing different OPC contents is shown in Fig. 4.31. The impact strength of the geopolymer matrix containing OPC is higher than that of pure geopolymer. Moreover, the enhancement in the impact strength for geopolymer matrix containing different OPC contents is like that of flexural strength. As indicated in Fig.4.31, the addition of OPC slightly enhanced the impact strength, with maximum improvement reaching 3.2 kJ/m^2 at 10 wt% OPC.

The impact strength of geopolymer increases with an increase in OPC content. The impact strength of neat geopolymer increased from 1.9 kJ/m^2 to 2.5, 2.9 and 3.2 kJ/m^2 with the addition of 5, 8 and 10 wt% of OPC, respectively. This enhancement in impact strength of the geopolymer matrix with increasing OPC content could be attributed to the OPC acting as crack stoppers and increasing the ability of the material to absorb energy by forming tortuous pathways for crack propagation which enhance the impact strength (Alamri and Low 2012a, b, c). These results confirm that the addition of OPC to a geopolymer system improves the mechanical properties of the matrix; they also suggest that an increase in OPC content is very useful in terms of improving the impact strength of a geopolymer matrix.

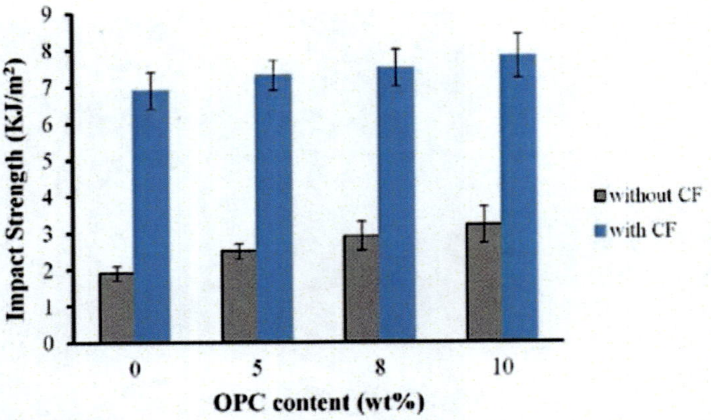

Fig. 4.31 Impact strength as a function of OPC content for geopolymers with and without CF

The impact strength of CF-reinforced geopolymer composites is also illustrated in Fig. 4.30. The cotton fibres significantly improved the impact strength of all composites, a function of their superior capacity to absorb impact energy than that of un-reinforced geopolymer. The addition of 5, 8 and 10 wt% OPC to CF-reinforced geopolymer was also found to improve the impact strength of the composites from 6.9 to 7.3 kJ/m^2, 7.5 and 7.8 kJ/m^2, respectively. This enhancement in impact strength is due to a good penetration of OPC into the cotton fabric, which strongly holds the filaments of the fabric together and leads to enhancement in the interfacial bonding between fabric and matrix (see Fig. 4.32). Fibre–matrix adhesion, as mentioned, is an important determinant of composite quality. In fibre-reinforced composites, the matrix has the role of transferring load to the fibres; this occurs through shear stresses at the interface. The effective transfer of load requires a strong bond to have formed between the polymeric matrix and the fibres (Chen et al. 2009). The presence of OPC plays an important role in improving the adhesion between fibre and matrix, thereby enhancing the mechanical properties of the composite.

Fracture toughness
The influence of OPC content on the fracture toughness of the geopolymer matrix and CF-reinforced geopolymer composites is shown in Fig. 4.33. The fracture toughness of geopolymer matrices containing 5 wt% OPC without CF fibres was lower than those of geopolymer composites containing 5 wt% OPC with CF fibres. The reason for such a difference is the ability of cotton fibres to resist fracture, which results in increased energy dissipation from crack-deflection at the fibre–matrix interface, fibre-debonding, fibre-bridging, fibre pull-out and fibre-fracture (Reis 2006; Silva et al. 2009, 2010; Tolêdo Filho et al. 2000, 2003). Previous studies have documented this improvement in fracture toughness in cement composites reinforced with natural fibres (Hakamy et al. 2013a, b; Sankar et al. 2017).

However, fracture toughness decreased when the OPC content increased beyond 5 wt%. It has already been considered that OPC addition results in desirable strength

Fig. 4.32 SEM observation of CF-reinforced geopolymer composite containing 10 wt% OPC

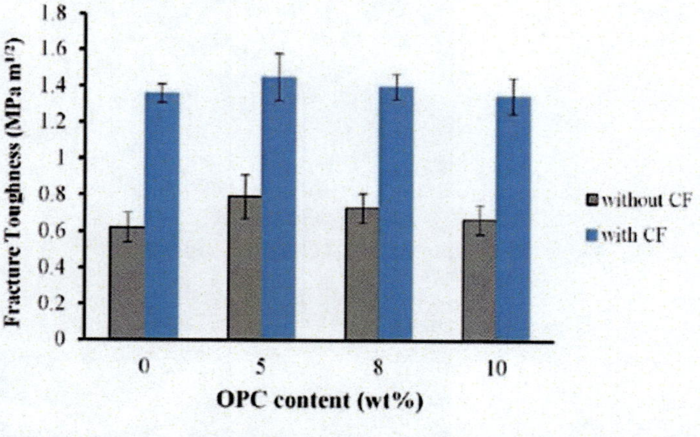

Fig. 4.33 Fracture toughness as a function of OPC content for geopolymers with and without CF

properties because of the improvement in fibre–matrix adhesion; but this also makes the composites brittle, as indicated by the lower fracture toughness results for all the samples. The addition of OPC causes fibre-matrix adhesion to be high but hinders the energy absorption mechanisms provided by fibre pull out and fibre de-bonding.

The fracture surfaces of all composites are shown in Fig. 4.34, which reveal an extensive occurrence of fibre pull-outs. These micrographs also reveal useful information about the fibre surfaces and interfacial debondings. It is worth noting that the pull-out lengths were greater in geopolymer composites with 5 wt% OPC

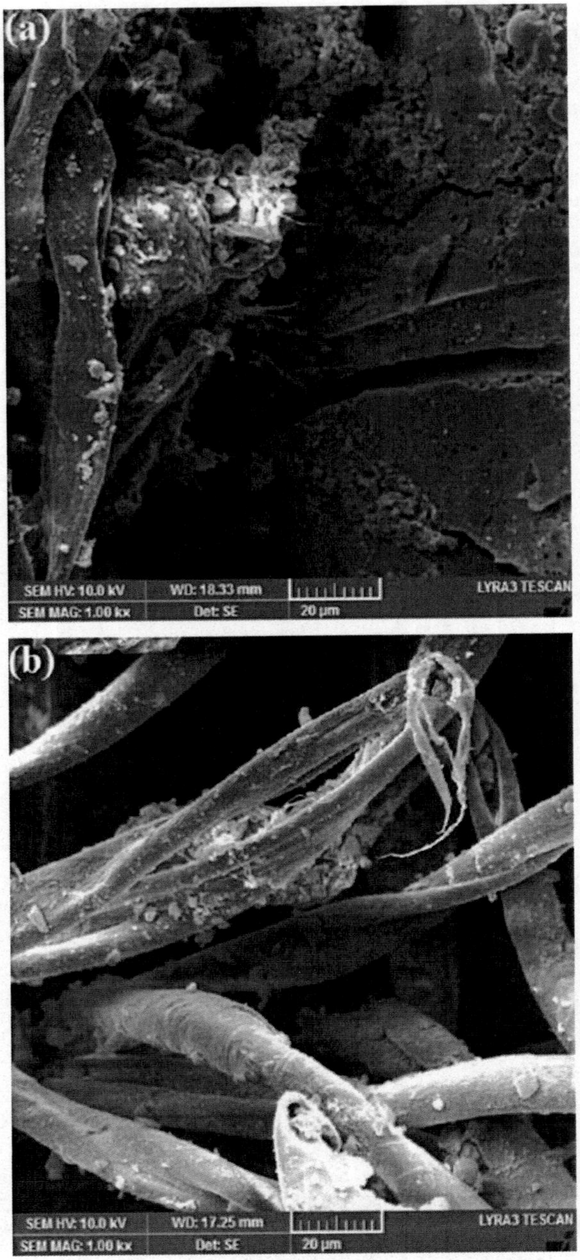

Fig. 4.34 SEM images showing the fracture surfaces of CF-reinforced geopolymer composite containing **a** 5 wt% OPC and **b** 10 wt% OPC

(Fig. 4.34a) than in composites with 10 wt% OPC (Fig. 4.34b). This supports the notion that the strength of interfacial adhesion is stronger in composites with 10 wt% OPC than in those with 5 wt% OPC. In the presence of weak interfacial adhesion, cracks tend to propagate through the fibre–matrix interfaces and result in greater fibre pull-out lengths. In contrast, when fibre–matrix adhesion is strong, propagation of cracks through the fibre–matrix interfaces is less, and thus short fibre pull-outs are observed. Interestingly, as the loading of OPC increased from 5 to 10 wt%, the lengths of fibre pull-outs reduced. For example, in the composites with 10 wt% OPC, almost no fibre pull-outs were observed, and the fibres were broken off with no interfacial debonding. These observations are the result of strong adhesion between the fibres and the matrix.

4.2.5 Effect of Elevated Temperature

Compressive strength

The influence of elevated temperature on compressive strength of geopolymer composites is exhibited in Fig. 4.35. The compressive strength of all geopolymer composites decreases after exposure to temperatures between 200 and 1000 °C. This reduction in compressive strength is probably because of the persistent deterioration of the geopolymer hydrates, which contributes most of the compressive strength of the composites, as temperature increases. Moreover, this decreasing tendency in compressive strength also results from increasing porosity as temperature increases. The total porosities of the composites heated to 800 and 1000 °C are higher than those at 200, 400 and 600 °C (see Fig. 6.3). The hydrophilic nature of cotton fibres enables the composites to take up moisture from the surrounding environment, increasing

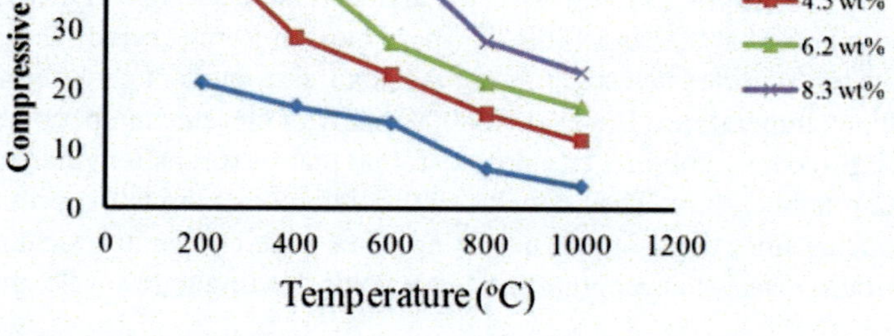

Fig. 4.35 Compressive strength of geopolymer composites at various temperatures

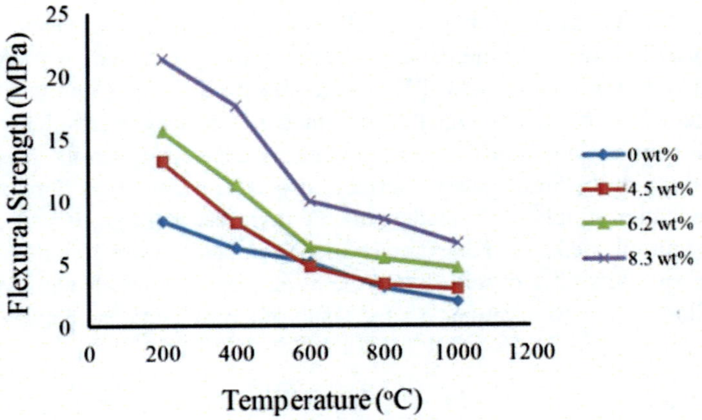

Fig. 4.36 Flexural strength of geopolymer composites at various temperatures

their water content. Upon heating, dehydration causes weight loss as would be expected. At low temperatures, dehydration is slow, but as temperature increases, so does the dehydration rate, leading to greater weight loss and the formation of a large quantity of voids that damage the bond between fibre and matrix and act as stress concentration points, resulting in a loss in load-bearing capacity (Low et al. 2007). Some authors carried out compressive tests on concrete containing polypropylene (PP) fibres and reported that, the strength properties decrease with the increase of temperature, due to the additional porosity and small channels created by the melting of the PP fibres (Noumowe 2005; Behnood and Ghandehari 2009; Uysal and Tanyildizi 2012).

Flexural strength

Like compressive strength, the flexural strength of all the composites decreases with an increase in temperature. As shown in Fig. 4.36, the reduction in flexural strength of the composites at 800 and 1000 is greater than those at 200, 400 and 600 °C. The main cause of this strength reduction may be attributed to fibre degradation, or to burning and void formation. When temperature increases, more fibres may degrade and more voids form, leading to a continual decrease in flexural strength. SEM examinations reveal some cotton fibres surviving inside the specimens heated between 200 and 600 (Fig. 4.37a–c): the possible reason for the higher flexural strength of these composites than those heated at 800 and 1000 °C where most of fibres degrade (Fig. 4.37d–e).

Of the few reported investigations of the flexural strength of geopolymer composites at high temperatures, Lin et al.'s (2009) study of elevated temperature on carbon fibre reinforced geopolymer composites found that flexural strength decreases with increasing temperature. They concluded that microcracking is the primary mechanism causing fibre degradation, occurring when high temperatures cause both free and hydration water to evaporate and leave voids, leading to lower flexural strength.

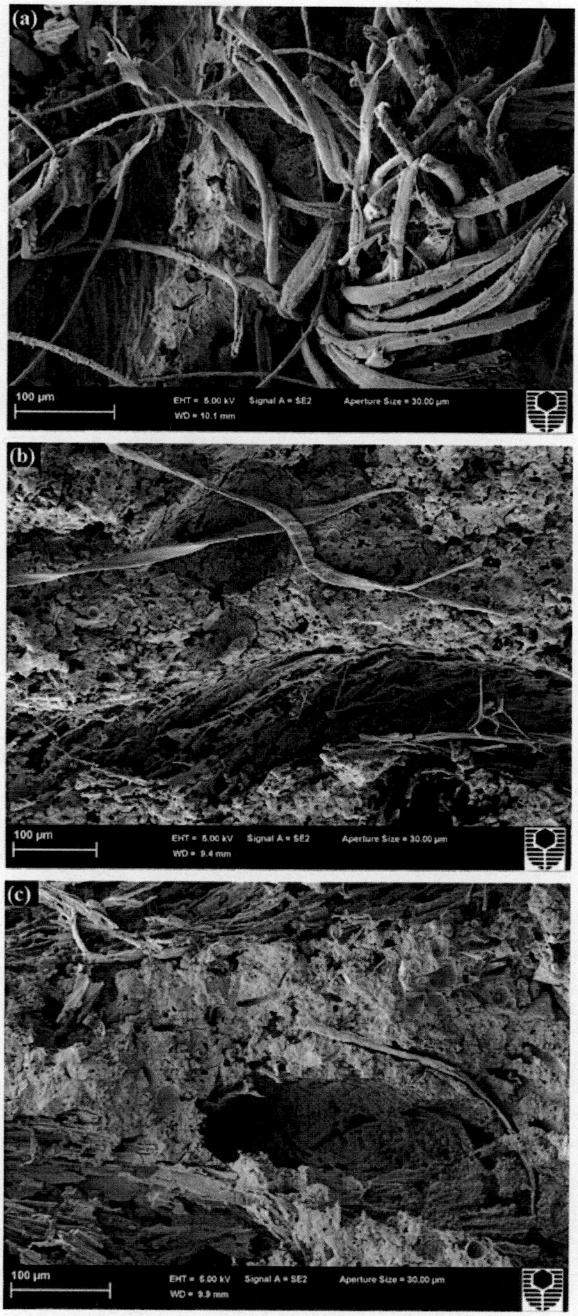

Fig. 4.37 SEM images of the fracture surface for geopolymer composites loaded with 8.3 wt% CF at various temperatures: **a** 200 °C, **b** 400 °C, **c** 600 °C, **d** 800 °C and **e** 1000 °C

Fig. 4.37 (continued)

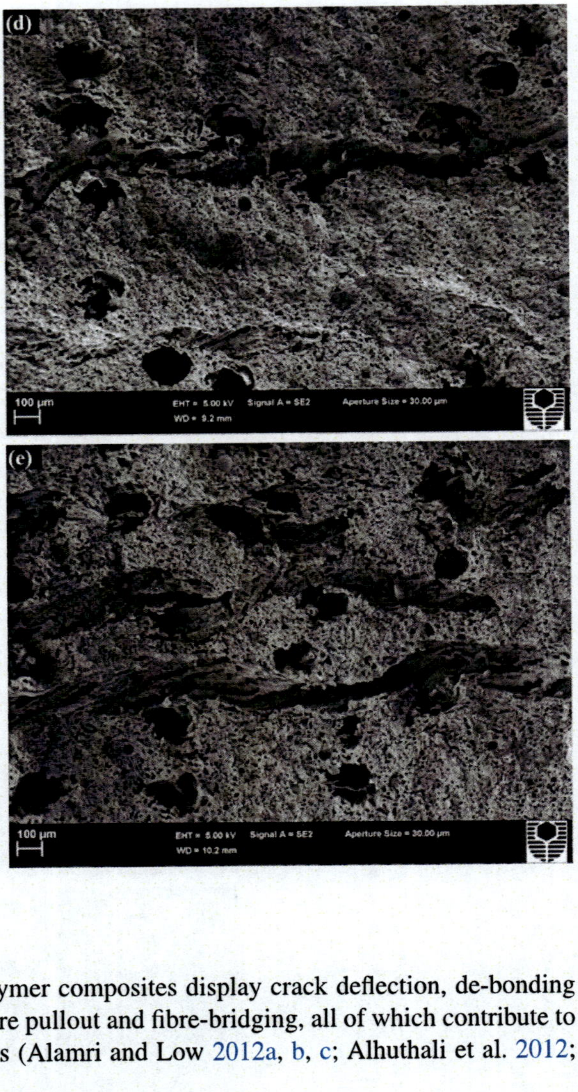

Fracture toughness

In general, natural fibre–polymer composites display crack deflection, de-bonding between fibre and matrix, fibre pullout and fibre-bridging, all of which contribute to improving fracture toughness (Alamri and Low 2012a, b, c; Alhuthali et al. 2012; Sankar et al. 2017).

Figure 4.38 shows that an increase in temperature of the geopolymer composites causes a decrease in fracture toughness. This can be explained by the very high porosity caused by the oxidation and consequent degradation of the cotton fibres, becoming more severe as temperature increases. At the highest temperatures tested (800 and 1000 °C), enhanced porosity formation in the matrices is observed, the probable result of the cotton fibre burning as shown in Fig. 4.37d–e. The images show high porosity in the composites as consequence of cotton fibre degradation due to oxidation effects, thus resulting in low fracture toughness because no fibre pull-out or fibre-fracture is observed. At testing temperatures below 800 °C, the

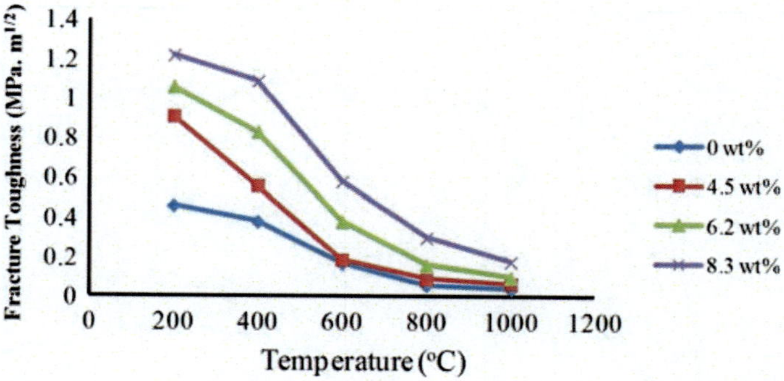

Fig. 4.38 Fracture toughness of geopolymer composites at various temperatures

higher fracture toughness of the composites may account for the presence of toughening mechanism such as fibre-bridging, fibre pull-out and fibre-fracture as only evident in the specimens exposed to 200–600 °C and confirmed by SEM observations (Fig. 4.38a–c).

4.3 Flax Fabric-Reinforced Geopolymer Composites

4.3.1 Effect of Flax-Fibre Content

Flexural strength and modulus

Generally, flexural tests are used to characterise the mechanical properties of layered composites as they provide a simple means of determining the bending response. This provides useful information on the performance of layered fabric-based composites (Abanilla et al. 2006). The effect of FF contents on the flexural stress–strain curves of the geopolymer composites is presented in Fig. 4.39. The composite containing 4.1 wt% FF showed the highest flexural strength among all composites. The flexural strength of the composites improved from 4.5 MPa in the pure geopolymer to about 23 MPa with 4.1 wt% FF. This result is comparable with that of short flax fibre-reinforced geopolymer composites reported by Alzeer and MacKenzie (2013). Both studies show that increasing the content of flax fibres leads to a significant improvement in the flexural strength of the composite. This can be explained by the fact that the number of reinforcement layers controls the flexural strength. The lower weight of flax fabrics allows multiple layers of fabric in the composite, to resist the shear failure and contribute in sustaining the applied load to the composites. This permits greater stress transfer between the matrix and the flax fibres, resulting in improved flexural strength (Sim et al. 2005; Korniejenko et al. 2018).

The flexural modulus of geopolymer composites, shown in Fig. 4.40, also indicates that the addition of FF to the matrix improves the flexural modulus over that of a pure

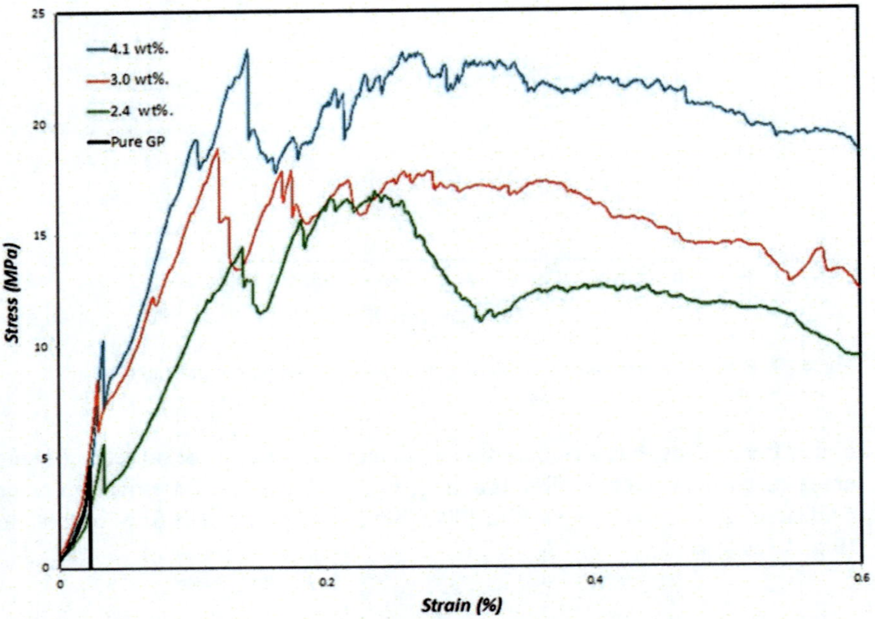

Fig. 4.39 Typical stress–strain curves of pure geopolymer and geopolymer composites with various FF contents

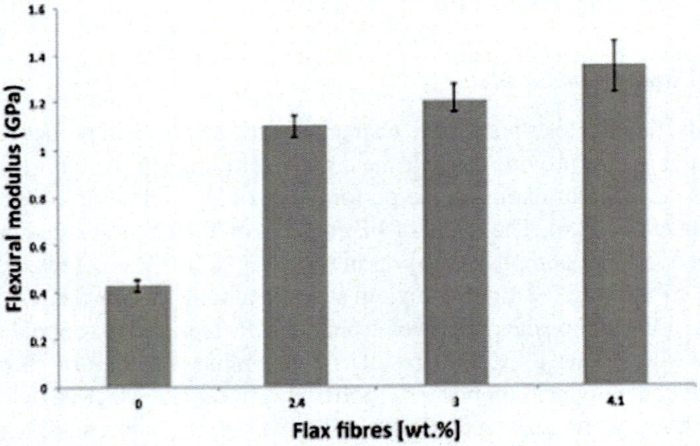

Fig. 4.40 Flexural modulus of geopolymer composites as a function of fabric content

geopolymer matrix. Flexural modulus is the measure of resistance to deformation of the composite in bending. It was observed that none of the reinforced specimens were completely broken at peak load. This could be attributed to crack bridging of the long continuous flax fibres under load, which makes the flexural modulus higher than that of pure geopolymer. Long fibres can withstand a higher load and are capable of

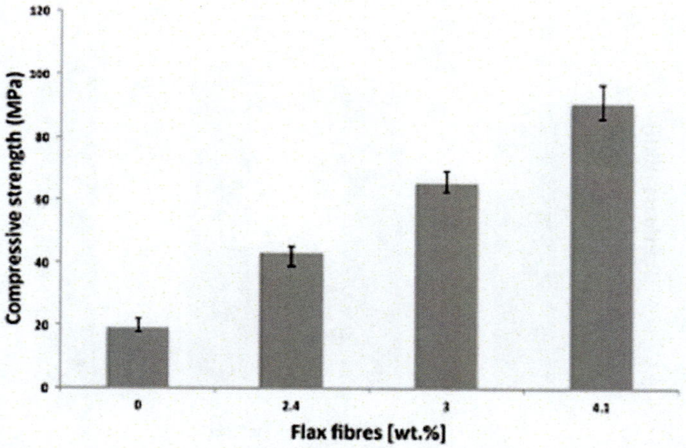

Fig. 4.41 Compressive strength of geopolymer composites as a function of fabric content

supporting multiple cracks throughout the loading process, consequently preventing brittle failure of the geopolymer.

Compressive strength

The results presented in Fig. 4.41 show that the compressive strength of the composites containing FF increases with increase in fibre contents. The increase in compressive strength with fibre loading may be due to the ability of the flax fibres to absorb stress transferred from the matrix. The compressive strength of the neat geopolymer paste increased from 19.4 to 91 MPa after the addition of 4.1 wt% flax fibres. This significant enhancement of compressive strength is because the interface between the fabric and the matrix is not exposed to any shear loading, which in turn reduces the possibility of fabric detachments or delamination from the matrix at high loads. Similar remarkable improvements in compressive strength have also been reported by Alomayri et al. (2014a, b) in the case of cotton fibre reinforced geopolymer composites. They concluded that the increase was due to the ability of horizontally laid cotton fabric to directly absorb and distribute a load uniformly throughout the cross-section.

Hardness

Hardness measurement enables the ability of a material to resist plastic deformation under indentation to be determined. The hardness values of FF-reinforced geopolymer composites are shown in Fig. 4.42. The results show that the hardness of composites increases with the addition of high number of flax fabrics to the geopolymer composite. This enhancement in hardness is due to the uniform distribution of the load on the flax fibres, which reduces the penetration of the test ball at the surface of the composite. A similar increase has been reported by other researchers studying natural fibre-reinforced geopolymer

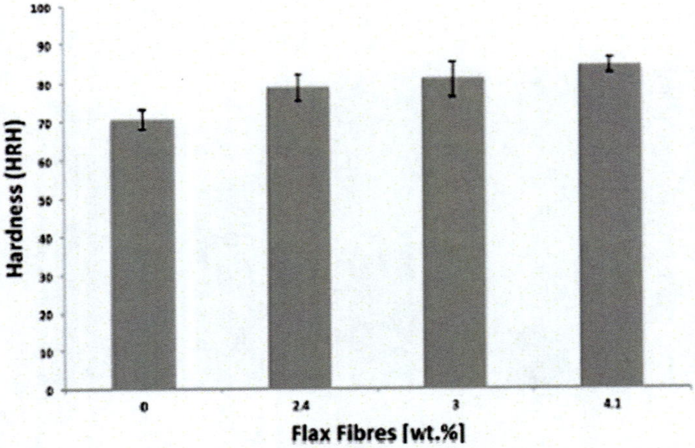

Fig. 4.42 Hardness of geopolymer composites as a function of fabric content

composites: for instance, Alomayri et al. (2014a, b) report that with increasing cotton fibre content, the hardness value of cotton fibre reinforced geopolymer composites increases (Alomayri et al. 2014a, b).

Fracture toughness

Generally, fibres' ability to resist crack deflection, debonding, and to bridge cracks, slows down crack propagation in fibre-reinforced composites and increases the fracture energy (Filho et al. 2003; Reis 2006; Silv et al. 2009, 2010). Figure 4.43 shows the influence of FF content on the fracture toughness of geopolymer composites. The composites containing FF show significantly higher fracture toughness than pure geopolymer matrix, and the higher the FF content the higher is the fracture

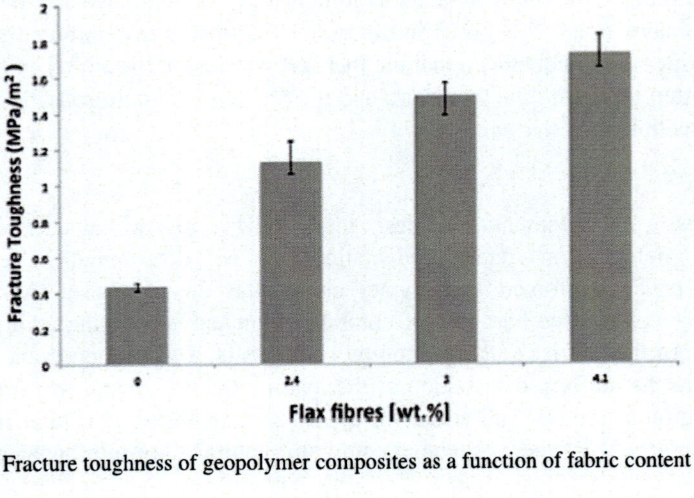

Fig. 4.43 Fracture toughness of geopolymer composites as a function of fabric content

toughness. The greatest improvement in fracture toughness was obtained from about 0.4 MPa m^{½} in the pure matrix to about 1.8 MPa m$^{1/2}$ with 4.1 wt% FF reinforcement. This extraordinary enhancement is due to the unique ability of flax fibre to resist fracture resulted in increased energy dissipation from crack-deflection at the fibre–matrix interface, fibre-debonding, fibre-bridging, fibre pull-out and fracture, clearly shown in the SEM images (see Fig. 4.44a–f). It can be seen in these images that small pieces of geopolymer paste attached to the fibre surface of the composites: such retention of the matrix on the fibre surfaces shows good adhesion between fibres and matrix.

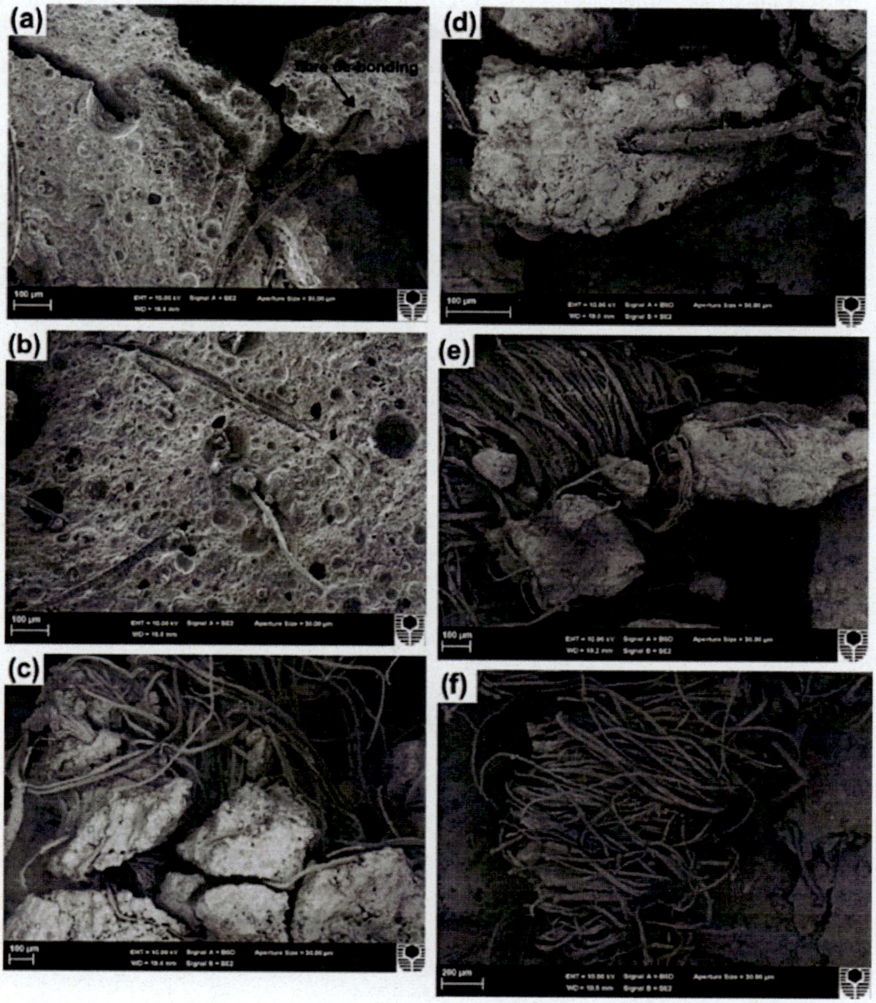

Fig. 4.44 SEM images of the fracture surface for geopolymer composites reinforced with flax fibres **a** shows fibre de-bonding, **b** fibre imprint and pull out, **c** fibre bridging cracks, **d** and **e** show the adhesion between fibre and matrix, and **f** fibre fracture

It was observed that the composites with fibres did not completely break into pieces, as the close spacing of woven FF led to crack-bridging by fibres and enhancing the resistance to their propagation. The effect of fibre content on the fracture surface can be seen by observing the difference between the matrix region and the fibre region. In Fig. 4.45a, b, composites filled with lower fibre content (2.4 and 3 wt%) show an increase in matrix-rich regions, which means there are insufficient fibres to transfer the load from the matrix. Due to this reason, the geopolymer composites with low fibre content exhibited low fracture toughness and mechanical properties. However, Fig. 4.45c illustrates the fracture surfaces of the geopolymer composites with higher fibre content, which means higher fibre-rich regions of composites with 4.1 wt% of FF. An increase in fibre-rich regions leads to greater stress-transfer from the matrix to the FF thereby resulting improvement of fracture toughness (Poletanovic et al. 2020; Simonova et al. 2018).

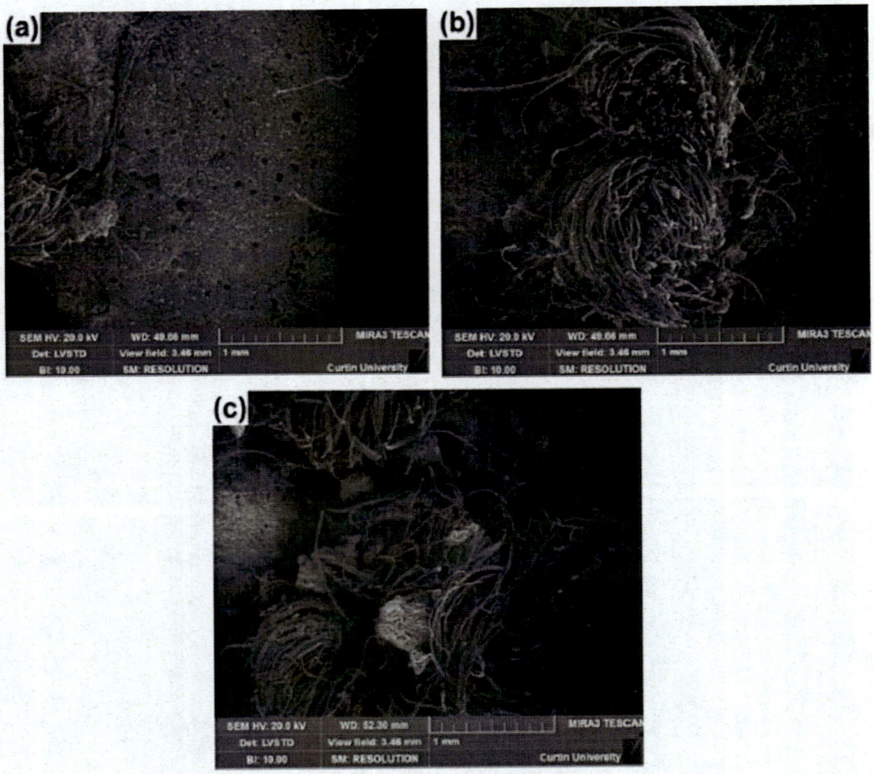

Fig. 4.45 Low magnification SEM images of the fracture surface for geopolymer composites reinforced with **a** 2.4, **b** 3.0 and **c** 4.1 wt% of flax fibres

4.3.2 Effect of Nanoclay

Mechanical properties

Flexural tests are used to characterise the mechanical properties of layered composites as they provide a simple means of determining the bending response. This provides useful information on the performance of layered fabric-based composites. The effect of nanoclay contents on the flexural strength of the geopolymer FF-composites is presented in Fig. 4.46. It can be seen clearly that all composites reinforced with FF showed higher flexural strength than the pure geopolymer and nanocomposites samples. The flexural strength of the composites improved from 4.5 MPa in the control sample to about 23 MPa in GPFNC-0. This result is comparable with that of short flax fibre-reinforced geopolymer composites reported by Alzeer and MacKenzie (2013). This can be explained by the fact that flax fabrics bridge the cracks of geopolymer matrix develop during bending and resisted the failure through frictional deboning of fabric in the matrix. This permits more stress transfer between the matrix and the flax fibres, resulting in greater flexural strength (Poletanovic et al. 2020; Sim et al. 2005).

The addition of nanoclay, however, enhanced the adhesion force between the matrix and fibres creating composites with higher flexural strength. Figure 4.46 shows that GPFNC-2 had the highest flexural strength among all samples, which means that the optimum addition that improved the flexural strength was 2.0 wt% of nanoclay. The loading of 2.0 wt% not only enhanced the bond between the matrix and the fibre, but also created a denser geopolymer paste with higher contents of

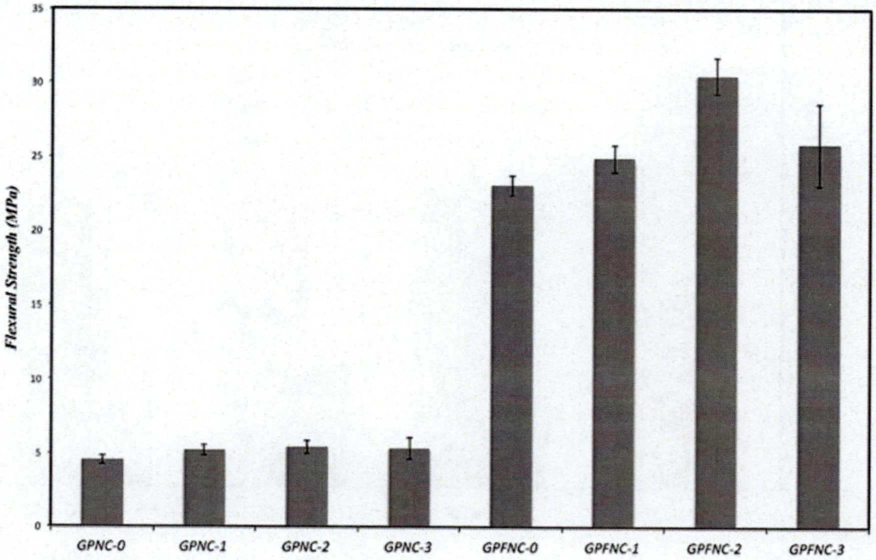

Fig. 4.46 Flexural strength of all samples

geopolymer products. This result is also confirmed by studying the flexural toughness indices I_5,I_{10} and $I_{failure}$ of the composites (Fig. 4.47a). According to the standard used, I_5 is defined as the ratio obtained by dividing the area up to a deflection of

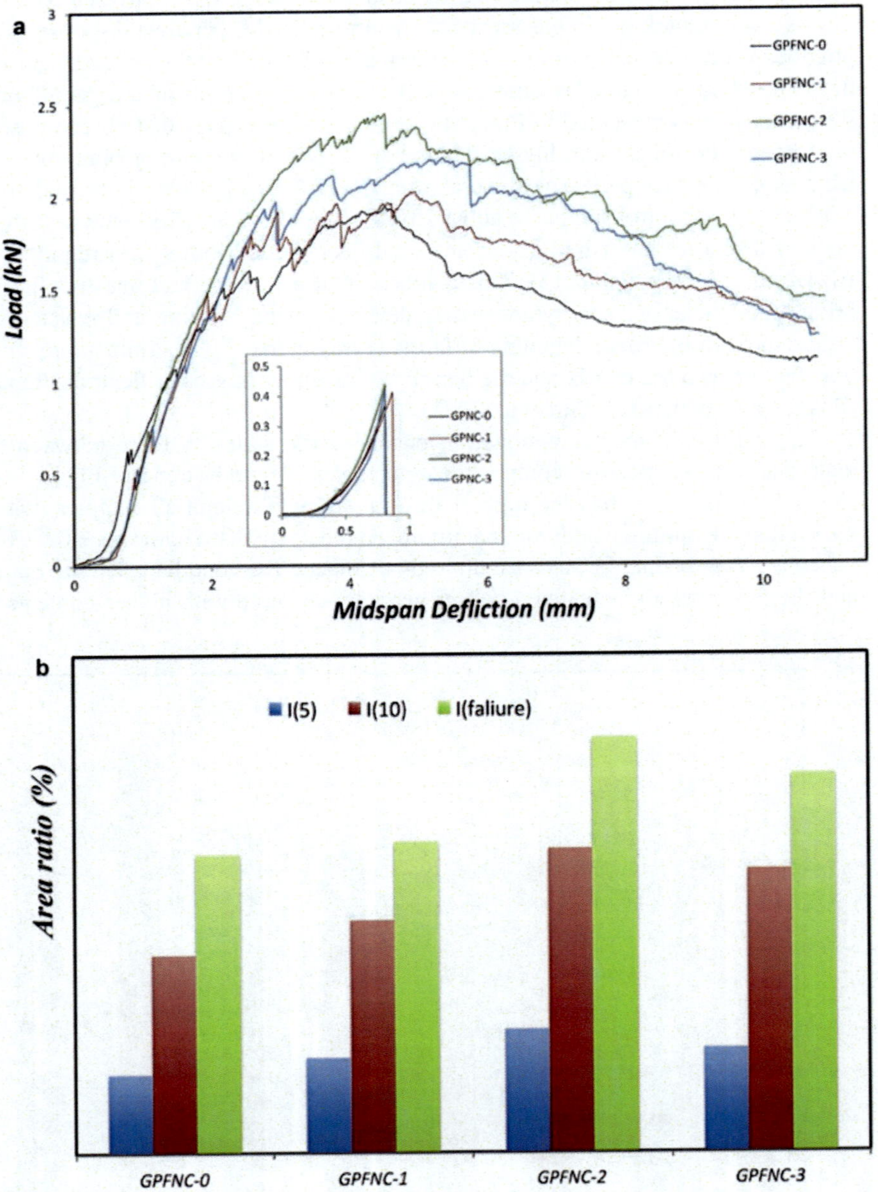

Fig. 4.47 **a** Typical load-midspan deflection curves of all composites, **b** Toughness indices I_5, I_{10} and $I_{failure}$ for FF/reinforced geopolymer samples

three times the first-crack deflection by the area up to first crack, while I_{10} is the ratio between the area up to a deflection of 5.5 times the first-crack deflection by the area up to the first crack. For the failure deflection, $I_{failure}$ is calculated at 11.4 mm deflection for all samples reinforced with FF.

Pure geopolymer and geopolymer nanocomposites had zero values of toughness because of the brittleness of the geopolymer. However, FF-reinforced composites exhibited high flexural toughness due to the ability of long fibres to withstand a higher load and to support multiple cracks throughout the loading process, which prevented the brittle failure of geopolymer.

Figure 4.47b presents values of toughness indices of FF-reinforced composites, the sample reinforced with the optimum loading of nanoclay showed higher toughness indices than GPFNC-0 by 58, 54 and 39% for I_5, I_{10} and $I_{failure}$, respectively. The rate of improvement of the toughness indices decreased with deflection. While I_5has enhanced by 58% after the addition of 2.0 wt% of nanoclay, $I_{failure}$ has only improved by 39%. This may be attributed to the effect of fibre pull-out that occurred more extensively in GPFNC-0 than in GPFNC-2. The bond between the matrix and flax fibres has improved due to the high content of geopolymer gel, which caused more fibres fracture than the pull-out in GPFNC-2. This can be considered clearly in Fig. 10a, where the slope of GPFNC-2 curve has sharper decrease in load with increasing deflection in the region between 9 and 11 mm than other curves.

SEM images of the fracture surface of FF-reinforced geopolymer composite and FF-reinforced nanocomposites after flexural toughness test are shown in Fig. 4.48. A range of toughness mechanisms such as fibre de-bonding, fibre pull-out and rupture and matrix fracture can be clearly seen. The examination of fracture surface of FF reinforced geopolymer composite shows high porous structure and number of unreacted fly ash, which caused poor adhesion between fibres and the matrix (Fig. 4.48a). FF-reinforced nanocomposites containing 1.0 and 3.0 wt% nanoclay displays relatively denser matrices with lower number of unreacted fly ash particles embedded in the matrices (Fig. 4.48b, d). However, in FF-reinforced geopolymer nanocomposite containing 2.0 wt% nanoclay, a smaller amount of unreacted fly ash particles was observed, and higher content of geopolymer gel can be clearly seen, which provided better adhesion between the flax fibres and the matrix. A significant amount of fibre fracture was also observed (Fig. 4.48c) by virtue of this enhanced interfacial fibre-matrix bonding.

4.3.3 *Effect of Nanosilica*

Flexural and compressive strengths

The effect of NS on the flexural and compressive strengths of geopolymer nanocomposites is presented in Fig. 4.49a, b. Overall, the incorporation of NS into geopolymer matrices led to noticeable enhancement in the flexural and compressive strength of geopolymer nanocomposites in the two mix approaches. The flexural strength of dry-mix nanocomposites containing 0.5, 1.0, 2.0 and 3.0 wt% NS is

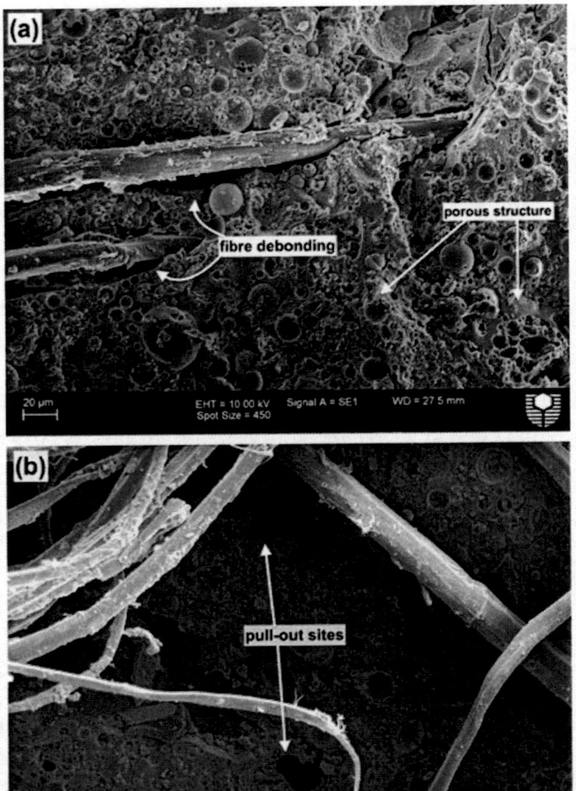

Fig. 4.48 SEM images of a fracture surface of FF-reinforced samples; **a** GPFNC-0, **b** GPFNC-1, **c** GPFNC-2, **d** GPFNC-3

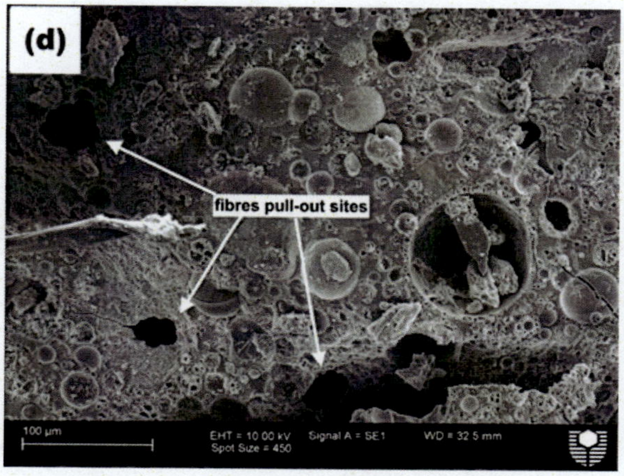

Fig. 4.48 (continued)

increased by 20.0, 28.8, 24.4 and 15.5%, respectively. While the flexural strength of wet-mix samples is increased by 8.8, 15.5, 22.2 and 13.3%, respectively, when compared to the control sample. The compressive strength results of geopolymer nanocomposites prepared by both dry and wet mix procedures imply comparable trends to the flexural strength results. The compressive strength of geopolymer paste improved from 37.2 to 47.3 and 44.9 MPa after the addition of 1.0 and 2.0 wt% NS in dry and wet mix samples, respectively. It can be noticed that the physical structure of geopolymer pastes has significant impacts on the mechanical behavior of geopolymer nanocomposites as the mechanical tests results followed similar trends to the densities of all samples. Flexural and compressive strengths are directly proportional to the nanocomposite's densities and inversely proportional to the porosities: denser specimens exhibited higher flexural and compressive strengths.

This improvement clearly indicates the efficiency of NS in improving the mechanical behavior in two ways: first, chemically by promoting the geopolymer reaction due to addition of nano silica to the system and forming additional sodium aluminosilicate hydrate or geopolymer gel (Phoo-ngernkham et al. 2014). This process occurs in both preparation methods; however, higher rates of silica was detected in the wet-mix process since the whole amounts of NS particles was dissolved in the alkaline solution, which accordingly produces geopolymer gel with higher content of silica. Secondly, NS particles could act as nano-fillers that fill the pores in geopolymer matrices (Qing et al. 2007). Therefore, the enhanced physical structure of geopolymer nanocomposite exhibited superior mechanical performance when compared to the control paste, particularly in the case of dry-mix procedure, where the nanoparticles could play both functions, filling the voids of the matrix beside the chemical role.

The flexural strength is related linearly to the square root of the compressive strength (Fig. 4.49c) in both dry-mix (Eq. 4.1) and wet-mix (Eq. 4.2) samples as follow:

$$\sigma_F = 1.65\sqrt{C} - 5.58 \tag{4.1}$$

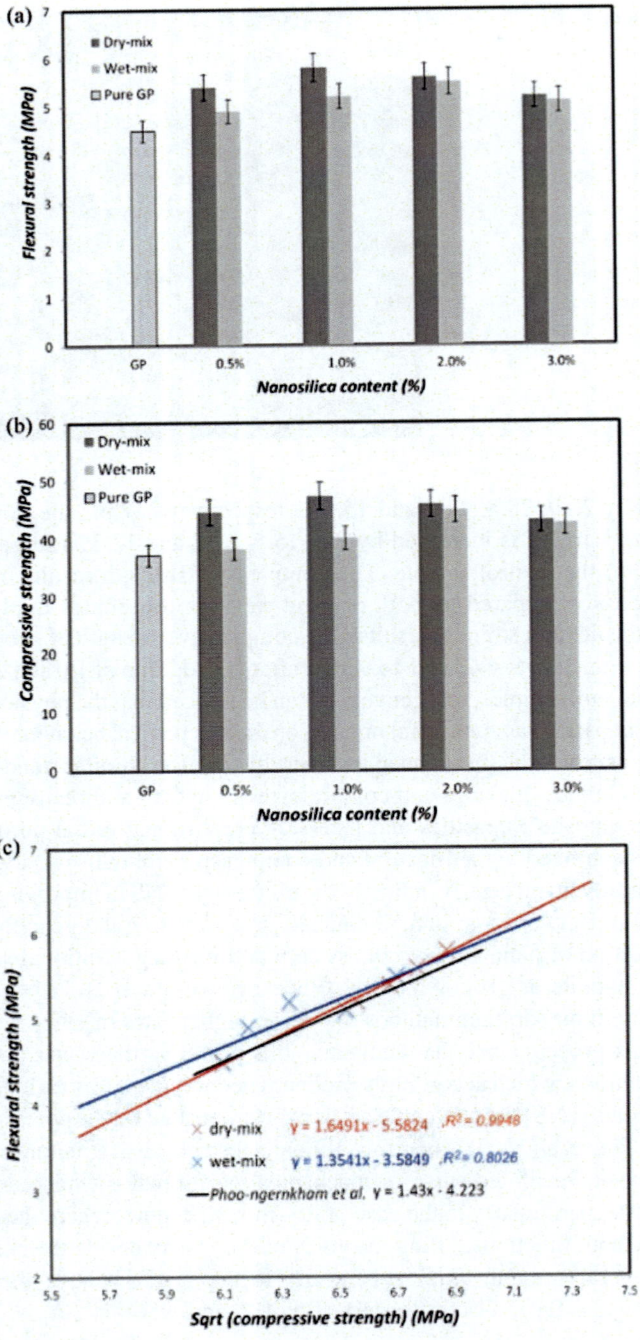

Fig. 4.49 Mechanical properties of geopolymer and geopolymer nanocomposites: **a** Flexural strength, **b** compressive strength, and **c** linear relationships between flexural strength and the square root of compressive strength of all samples

$$\sigma_F = 1.35\sqrt{C} - 3.58 \quad (4.2)$$

where σ_F is the flexural strength and C is the compressive strength of the samples. A similar trend that represents similar relationship of NS-reinforced geopolymer composite reported by Phoo-ngernkham et al. (2014) was also shown in the same figure for comparison and indicate comparable results. They found that the addition of 2.0 wt% NS to geopolymer pastes has improved the flexural and compressive strengths by 44.8% and 31.4%, respectively. In another study, Qing et al. (2007) reported the influence of 3.0 wt% nano-SiO_2 addition on the properties of cement paste and observed that the flexural strength increased by about 72% compared to control cement matrix. The results were attributed this enhancement to the pozzolanic and filler effects of nano-SiO_2 particles. In a further research, the influence of NS on the mechanical properties of cement mortar at 28 days with water/binder ratio of 0.4 has been studied (Hosseini et al. 2014). It has been found that the addition of 1.0 wt% NS improved the flexural strength from 7.0 to 9.3 MPa, and the compressive strength from 50.1 to 56.7 MPa, about 33 and 13% increase, respectively. Furthermore, the effect of nanoclay (Cloisite 30B) on the physical and mechanical properties of fly ash based geopolymer has been reported in previous study (Assaedi et al. 2016), and found that 2.0 wt% loading of nanoclay has improved the flexural and compressive strength by 20% and 23%, respectively.

The trends, however, are reversed after addition of high amounts of NS in both cases. The reduction in mechanical properties of wet-mix samples containing high amount of NS is due to the excessive amounts of silica dissolved in the system, which led to an inadequate OH^- ions that fully dissolve the aluminum ions (Al^{3+}), leaving unreacted flyash in the sample (Rowles and O'Connor 2003). While the reduction in the mechanical properties of dry-mix samples containing high amount of NS could be attributed to the relatively poor dispersion and agglomerations of NS particles in geopolymer matrices at higher NS contents, which create weak zones in the form of micro-pores (Naji Givi et al. 2010; Hakamy et al. 2014; Assaedi et al. 2016). Nevertheless, the addition of NS improved the mechanical strength of geopolymer nanocomposite when compared to the control geopolymer paste regardless of the mixing method and the amount of NS added to the pastes. Though the flexural and compressive strengths of geopolymer nanocomposite with high content of NS are decreased compared to the nanocomposite with the optimum loading in dry/wet-mixing approaches, they are still higher than the pure geopolymer matrix.

Flexural toughness indices

The flexural toughness of FF-reinforced nanocomposites was characterised and evaluated by the toughness indices I_5, I_{10} and $I_{failure}$ as defined by ASTM C1018, Standard Test Method for Flexural Toughness and First-Crack Strength of Fiber-Reinforced Concrete (Using Beam with Third-Point Loading). According to the standard, I_5 is defined as the ratio obtained by dividing the area up to a deflection of three times the first-crack deflection by the area up to first crack deflection, while I_{10} is the ratio of the area up to a deflection of 5.5 times the first-crack deflection to the area up to the first crack (see Fig. 4.50). Thus, it could be written as:

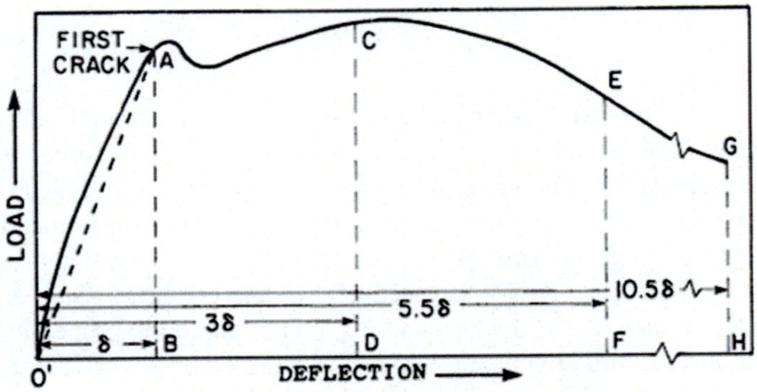

Fig. 4.50 Definition of ASTM toughness indices [ASTM C1018]

$$I_5 = \frac{(Area)_{OACDO}}{(Area)_{OABO}} \tag{4.3}$$

$$I_{10} = \frac{(Area)_{OAEFO}}{(Area)_{OABO}} \tag{4.4}$$

Similarly, $I_{failure}$ can be evaluated for the failure deflection of 11.4 mm in this study as most of the specimens lost their load carrying capacity significantly at that deflection.

Pure Geopolymer matrices and geopolymer nanocomposite are brittle in nature and crack easily under applied forces; thus, they do not exhibit any toughness. To improve the capability of load capacity at strains greater than that at the initial crack, it is essential to reinforce the matrix using fibres such as FF. In previous study, pure geopolymer pastes have been reinforced with various weight contents of FF, which exhibited higher toughness as the natural fibres content increased (Simonova et al. 2018; Assaedi et al. 2015). Composites reinforced with high amount of FF demonstrated greater flexural toughness because of the capability of FF to bear the higher loads and to support the multiple cracks during the loading process, which avoided the brittle failure of geopolymers. In the current study, flexural toughness of hybrid composites containing a combination of NS and FF are studied.

The ability of FF-reinforced geopolymer nanocomposites to absorb energy is identified by studying the flexural toughness indices I_5, I_{10} and $I_{failure}$. For the failure deflection, $I_{failure}$ is calculated at 11.4 mm deflection for geopolymer composites and nanocomposites reinforced with FF. Plots in Fig. 4.51a, b display the load versus midspan deflection curves of FF-reinforced geopolymer composites and nanocomposites. FF-reinforced nanocomposite with the optimum loading of NS in both groups showed likewise optimum flexural toughness. Figure 4.51c presents the toughness indices of FF-reinforced composites. In the dry-mix samples, the FF-reinforced nanocomposite loaded with 1.0 wt% showed the highest toughness indices of 44, 125 and 146 while the toughness indices of wet-mix sample loaded with 2.0 wt% NS were 43, 114 and 133 for I_5, I_{10} and $I_{failure}$, respectively. This improvement could be

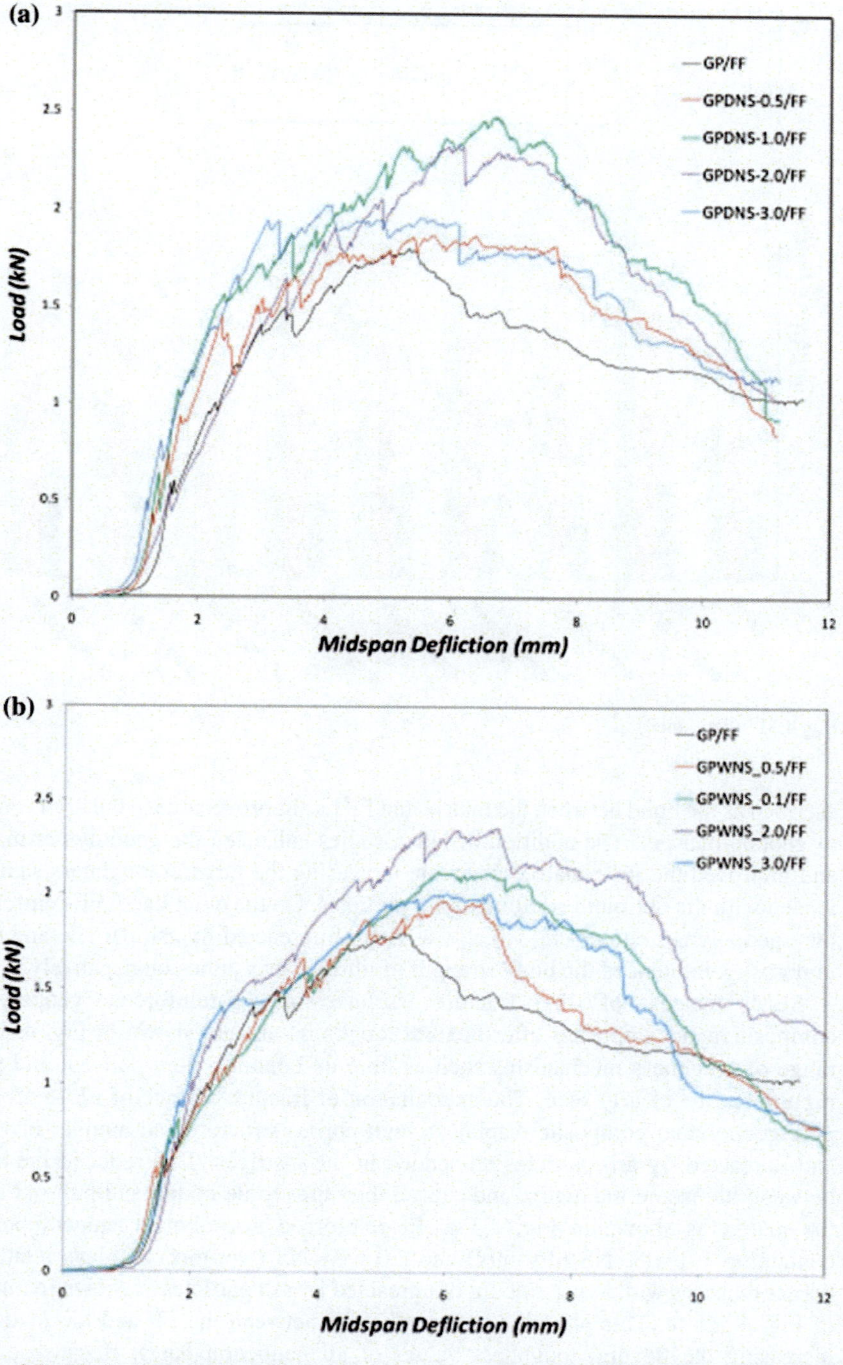

Fig. 4.51 Typical load-midspan deflection curves of: **a** FF-reinforced geopolymer nanocomposites prepared by dry-mix approach, **b** FF-reinforced geopolymer nanocomposites prepared by wet-mix approach, **c** Toughness indices I_5, I_{10} and I_{failure} for all FF-reinforced geopolymer samples

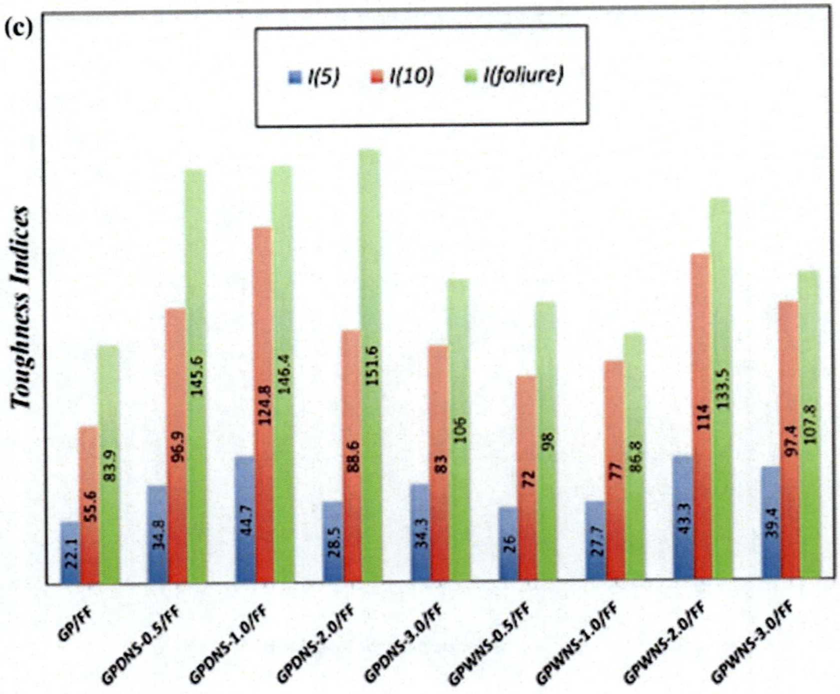

Fig. 4.51 (continued)

ascribed as the bond between the matrix and FF has improved due to the high content of geopolymer gel. The addition of NS particles enhanced the geopolymer matrix and improved the fibre-matrix adhesion, increasing the flexural toughness samples loaded with the optimum addition of NS particles. On the other hand, FF-reinforced pure geopolymer composite, which was rich of unreacted fly ash particle and high in porosity, influenced the bond strength of fibre–matrix adhesion negatively.

SEM images of the fracture surface of FF-reinforced geopolymer composite/nanocomposites after flexural toughness test are shown in Fig. 4.52. A range of toughness mechanisms such as fibre de-bonding, fibre pull-out and fibre rupture can be clearly seen. The examination of fracture surface of FF reinforced pure geopolymer composite displays a high porous structure and number of voids and unreacted fly ash particles embedded in the matrices. This reduced the bond between fibres and the matrix and caused the fibres to de-bound and pull out from the matrix as shown in Fig. 4.52a. FF-reinforced geopolymer nanocomposites containing 1.0 wt% NS (dry-mix) and 2.0 wt% NS (wet-mix) displays relatively denser matrices with lower amount of unreacted fly ash particles as shown in images of Fig. 4.52b, c. This provided better adhesion between the FF and the matrices, increasing the flexural toughness values of all nanocomposites. Because of the strong fibre-matrices bonds, fibres fracture has been seen in the fracture surfaces of the nanocomposites.

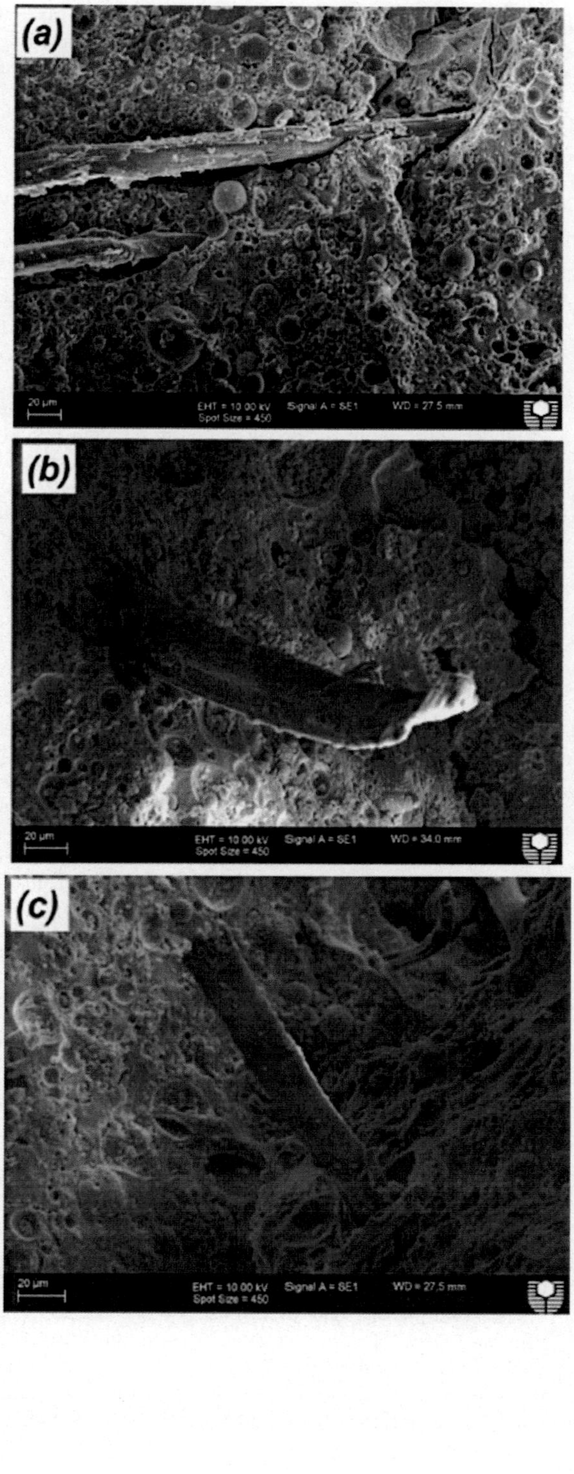

Fig. 4.52 SEM images showing the fracture surfaces of FF-reinforced samples; **a** GP/FF, **b** GPDNS-1/FF, and **c** GPWNS/FF-2

Acknowledgements The authors would like to thank Ms. E. Miller for assistance with SEM. The authors also thank Mr. Andreas Viereckle of Mechanical Engineering for assistance with Charpy impact test. The collection of diffraction data was funded by the Australian Synchrotron (Proposal FI5075).

References

Abanilla MA, Karbhari VM, Li Y (2006) Interlaminar and interlaminar durability characterization of wet layup carbon/epoxy used in external strengthening. Compos Part B 37:650–661

Alamri H, Low IM (2012) Characterization of epoxy hybrid composites filled with cellulose fibres and nano-SiC. J Appl Polym Sci 126:221–231

Alamri H, Low IM (2012) Mechanical properties and water absorption behaviour of recycled cellulose fibre reinforced epoxy composites. Polym Test 31:620–628

Alamri H, Low IM (2012) Microstructural, mechanical, and thermal characteristics of recycled cellulose fiber-halloysite-epoxy hybrid nanocomposites. Polym Compos 33:589–600

Alhuthali A, Low IM, Dong C (2012) Characterisation of the water absorption, mechanical and thermal properties of recycled cellulose fibre reinforced vinyl-ester eco-nanocomposites. Compos B 43:2772–2781

Alomayri T, Low IM (2013) Synthesis and characterization of mechanical properties in cotton fiber-reinforced geopolymer composites. J Asian Ceram Soc 1:30–34

Alomayri T, Shaikh FUA, Low IM (2013) Characterisation of cotton fibre-reinforced geopolymer composites. Compos Part B Eng 50:1–6

Alomayri T, Shaikh FUA, Low IM (2013) Thermal and mechanical properties of cotton fabric-reinforced geopolymer composites. J Mater Sci 48:6746–6752

Alomayri T, Shaikh FUA, Low IM (2014) Effect of fabric orientation on mechanical properties of cotton fabric reinforced geopolymer composites. Mater Des 57:360–365

Alomayri T, Shaikh FUA, Low IM (2014) Synthesis and mechanical properties of cotton fabric reinforced geopolymer composites. Compos Part B 60:36–42

Alzeer M, MacKenzie K (2013) Synthesis and mechanical properties of novel composites of inorganic polymers (geopolymers) with unidirectional natural flax fibres (phormium tenax). Appl Clay Sci 75:148–152

Assaedi H, Alomayri T, Shaikh FUA, Low IM (2015) Characterisation of mechanical and thermal properties in flax fabric reinforced geopolymer composites. J Adv Ceram 4:272–281

Assaedi H, Shaikh FUA, Low IM (2016) Effect of nano-clay on mechanical and thermal properties of geopolymer. J Asian Ceram Soc 4:19–28

Avella M, Buzarovska A, Errico ME, Gentile G, Grozdanov A (2009) Eco-challenges of bio-based polymer composites. Materials 2:911–925

Behnood A, Ghandehari M (2009) Comparison of compressive and splitting tensile strength of high-strength concrete with and without polypropylene fibers heated to high temperatures. Fire Safety J 44:1015–1022

Boonserma K, Sata V, Pimraksab K, Chindaprasirta P (2012) Improved geopolymerization of bottom ash by incorporating fly ash and using waste gypsum as additive. Cem Concr Compos 34:819–824

Chen H, Miao M, Ding X (2009) Influence of moisture absorption on the interfacial strength of bamboo/vinyl ester composites. Compos A Appl Sci Manuf 40:2013–2019

Dias DP, Thaumaturgo C (2005) Fracture toughness of geopolymeric concretes reinforced with basalt fibres. Cem Concr Compos 27:49–54

Dlugosz J (1965) The fine structure of cotton fibre as revealed by swelling during methacrylate embedding. Polymer 6:427–436

Dombrowski K, Buchwald A, Weil M (2007) The influence of calcium content on the structure and thermal performance of fly ash based geopolymers. J Mater Sci 42:3033–3043

Filho RDT, Ghavami K, England GL (2003) Development of vegetable fibre-mortar composites of improved durability. Cem Concr Compos 25:185–196

Franco-Marquès E, Méndez JA, Pèlach MA, Vilaseca F, Bayer J, Mutjé P (2011) Influence of coupling agents in the preparation of polypropylene composites reinforced with recycled fibers. Chem Eng J 166:1170–1178

Graupner N (2008) Application of lignin as natural adhesion promoter in cotton fibre-reinforced poly(lactic acid) (PLA) composites. J Mater Sci 43:5222–5229

Hakamy A, Shaikh FUA, Low IM (2013) Microstructures and mechanical properties of hemp fabric reinforced organoclay–cement nanocomposites. Constr Build Mater 49:298–307

Hakamy A, Shaikh FUA, Low IM (2013) Thermal and mechanical properties of hemp fabric-reinforced nanoclay–cement nanocomposites. J Mater Sci 49:1684–1694

Hakamy A, Shaikh FUA, Low IM (2014) Thermal and mechanical properties of hemp fabric-reinforced nanoclay–cement nanocomposites. J Mater Sci 49:1684–1694

Hosseini P, Hosseinpourpia R, Pajum A, Khodavirdi MM, Izadi H, Vaezi A (2014) Effect of nanoparticles and aminosilane interaction on the performances of cement-based composites: an experimental study. Constr Build Mater 66:113–124

Joseph PV, Joseph K, Thomas S (1999) Effect of processing variables on the mechanical properties of sisal fiber-reinforced polypropylene composites. Compos Sci Technol 59:1625–1640

Khan JA, Khan MA, Islam R (2012) Effect of potassium permanganate on mechanical thermal and degradation characteristics of jute fabric-reinforced polypropylene composite. J Reinf Plast Compos 31:1725–1736

Korniejenko K, Łach M, Hebdowska-Krupa M, Mikuła J (2018) The mechanical properties of flax and hemp fibres reinforced geopolymer composites. IOP Conf Ser Mater Sci Eng 379, 012023

Li W, Xu J (2009) Mechanical properties of basalt fibre reinforced geopolymeric concrete under impact loading. Mater Sci Eng 505:178–186

Lin T, Jia D, He P et al (2009) Thermal-mechanical properties of short carbon fiber reinforced geopolymer matrix composites subjected to thermal load. J Cent South Univ Tech 16:881–886

Low IM, McGrath M, Lawrence D, Schmidt P, Lane J, Latella BA et al (2007) Mechanical and fracture properties of cellulose-fibre-reinforced epoxy laminates. Compos Part A Appl Sci Manuf 38:963–974

Low IM, Somers J, Kho HS, Davies IJ, Latella BA (2009) Fabrication and properties of recycled cellulose fibre-reinforced epoxy composites. Compos Interfac 16:659–669

Mishra S, Mohanty AK, Drzal LT, Misra M, Parija S, Nayak SK, Tripathy SS (2003) Studies on mechanical performance of biofibre/glass reinforced polyester hybrid composites. Compos Sci Technol 63:1377–1385

Mobasher B, Peled A, Pahilajani J (2006) Distributed cracking and stiffness degradation in fabric-cement composites. Mater Struct 39:317–331

Naji Givi A, Abdul Rashid S, Aziz FNA, Salleh MAM (2010) Experimental investigation of the size effects of SiO_2 nanoparticles on the mechanical properties of binary blended concrete. Compos Part B Eng 41:673–677

Nakagaito AN, Yano H (2004) The effect of morphological changes from pulp fibre towards nano-scale fibrillated cellulose on the mechanical properties of high-strength plant fibre based composites. Appl Phys A 78:547–552

Noumowe A (2005) Mechanical properties and microstructure of high strength concrete containing polypropylene fibres exposed to temperatures up to 200 °C. Cement Concr Res 35:2192–2198

Pakravan H, Jamshidi M, Latifi M, Neshastehriz M (2011) Application of polypropylene nonwoven fabrics for cement composites reinforcement. Asian J Civ Eng 12:551–562

Phair JW, van Deventer JSJ (2001) Effect of silicate activator pH on the leaching and material characteristics of waste-based inorganic polymers. Miner Eng 14:289–304

Phoo-ngernkham T, Chindaprasirt P, Sata V, Hanjitsuwan S, Hatanaka S (2014) The effect of adding nano-SiO_2 and nano-Al_2O_3 on properties of high calcium fly ash geopolymer cured at ambient temperature. Mater Des 55:58–65

Poletanovic B, Dragas J, Ignjatovic I, Komljenovic M, Merta I (2020) Physical and mechanical properties of hemp fibre reinforced alkali-activated fly ash and fly ash/slag mortars. Constr Build Mater 259:119677

Qing Y, Zenan Z, Deyu K, Rongshen C (2007) Influence of nano-SiO_2 addition on properties of hardened cement paste as compared with silica fume. Constr Build Mater 21:539–545

Reis JML (2006) Fracture and flexural characterization of natural fiber-reinforced polymer concrete. Constr Build Mater 20:673–678

Rickard WDA, Williams R, Temuujin J, van Riessen A (2011) Assessing the suitability of three Australian fly ashes as an aluminosilicate source for geopolymers in high temperature applications. Mater Sci Eng 528:3390–3397

Rowles M, O'Connor B (2003) Chemical optimisation of the compressive strength of aluminosilicate geopolymers synthesised by sodium silicate activation of metakaolinite. J Mater Chem 13:1161–1165

Sankar K, Sá Ribeiro RA, Sá Ribeiro MG, Kriven WM (2017) Potassium-based geopolymer composites reinforced with chopped bamboo fibers. J Am Ceram Soc 100:49–55. https://doi.org/10.1111/jace.14542

Silva FA, Mobasher B, Tolêdo Filho RD (2009) Cracking mechanisms in durable sisal fibre reinforced cement composites. Cem Concr Compos 31:721–730

Silva FA, Filho RDT, Filho JAM, Fairbairn EMR (2010) Physical and mechanical properties of durable sisal fibre–cement composites. Constr Build Mater 24:777–785

Sim J, Park C, Moon DY (2005) Characteristics of basalt fiber as a strengthening material for concrete structures. Compos Part B 36:504–512

Simonova H, Kucharczykova B, Topolar L, Kersner Z, Merta I, Dragas J, Ignjatovic I, Komljenovic M, Nikolic V (2018) Crack initiation of selected geopolymer mortars with hemp fibers. Procedia Struct Integr 13:578–583

Soroushian P, Elzafraney M, Nossoni A, Chowdhury H (2006) Evaluation of normal weight and light-weight fillers in extruded cellulose fiber cement products. Cem Concr Compos 28:69–76

Talimi M, Rizvi G (2008) Properties enhancement of PLA-natural fibre composites using an ethylene copolymer. World J Eng 20:1461–1462

Tolêdo Filho RD, Romildo D, Scrivener K, England GL, Ghavami K (2000) Durability of alkali-sensitive sisal and coconut fibres in cement mortar composites. Cem Concr Compos 22:127–143

Tolêdo Filho RD, Romildo D, Khosrow G, England GL, Scrivener K (2003) Development of vegetable fibre–mortar composites of improved durability. Cem Concr Compos 25:185–196

Uysal M, Tanyildizi H (2012) Estimation of compressive strength of self compacting concrete containing polypropylene fiber and mineral additives exposed to high temperature using artificial neural network. Constr Build Mater 27:404–414

van Jaarsveld JGS, van Deventer JSJ, Schwartzman A (1999) The potential use of geopolymeric materials to immobilise toxic metals. Material and leaching characteristics. Miner Eng 12:75–91

Wambua P, Ivens J, Verpoest I (2003) Natural fibres: Can they replace glass in fibre reinforced plastics? Compos Sci Technol 63:1259–1264

Yip CK, van Deventer JSJ (2003) Microanalysis of calcium silicate hydrate gel formed within a geopolymeric binder. J Mater Sci 38:3851–3860

Yip CK, Lukey GC, Van Deventer JSJ (2005) The coexistence of geopolymeric gel and calcium silicate hydrates at the early stage of alkaline activation. Cem Concr Res 35:1688–1697

Zhang J, Li VC (2004) Simulation of crack propagation in fibre-reinforced concrete by fracture mechanics. Cem Concr Res 34:333–339

Zhang W, Okubayashi S, Bechtold T (2005) Fibrillation tendency of cellulosic fibers part 3: effects of alkali pretreatment of lyocell fiber. Carbohydr Polym 59:173–179

Zhang Y, Sun W, Li Z (2006) Impact behavior and microstructural characteristics of PVA fiber reinforced fly ash-geopolymer boards prepared by extrusion technique. J Mater Sci 41:2787–2794

Chapter 5
Moisture Absorption and Durability

Abstract This chapter describes the characteristics of moisture absorption and durabilty of geopolymer reinforced with natural fibres and/or nanofillers. Results show that the appropriate addition of natural fibres and/or nanofillers can have a profound influence on the moisture resistance of geopolymer composites and nanocomposites. The durability of these materials was ascertained by studying the effect of water absorption on their mechanical and physical properties.

5.1 Cotton Fabric-Reinforced Geopolymer Composites

5.1.1 Water Absorption Behaviour

Figure 5.1 shows the percentage of water uptake as a function of square root of time of geopolymer composite samples reinforced with 0, 4.5, 6.2 and 8.3 wt% CF due to immersion in tap water for 133 days at room temperature. The water absorption increases with increase in fibre contents. The increase in water absorption is due to the hydrophilic nature of natural fibre and the greater interfacial area between the fibre and the matrix (Dhakal et al. 2007). The maximum water uptake and the diffusion coefficient values increased for all composite specimens as the cotton fibre content increased (see Table 5.1). The water absorption of all specimens was high in the early stages of exposure, after which it slowed down and reached saturation level after prolonged time, following a Fickian diffusion process. The initial rate of water absorption and the maximum water uptake increase as the fibre loading increases in all-natural fibre composite samples (Dhakal et al. 2007). This phenomenon can be explained by considering the water uptake characteristics of cotton fibre. When natural fibre-reinforced composite is exposed to moisture, the hydrophilic nature of fibre, in this case cotton, causes the fibre to absorb water and swell. As a result, micro-cracking of the geopolymer composite occurs. The high cellulose content in cotton fibre absorbs extra water that penetrates the interface through these micro-cracks, creating swelling stresses that lead to composite failure (Graupner 2008). The more the composite cracks, the more capillarity and transport via micro cracks become

I.-M. Low et al., *Cotton and Flax Fibre-Reinforced Geopolymer Composites*, Composites Science and Technology,
https://doi.org/10.1007/978-981-16-2281-6_5

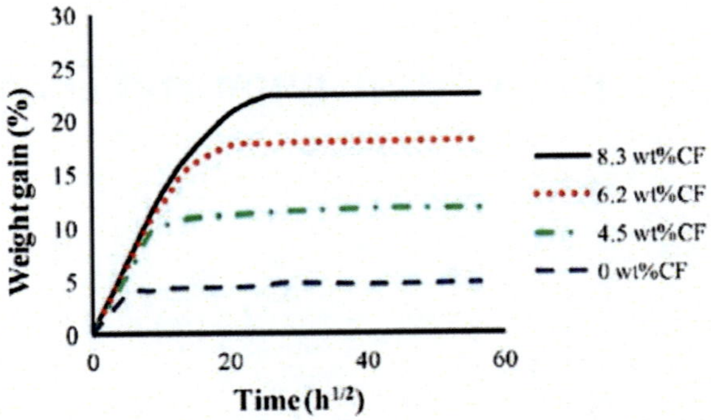

Fig. 5.1 Water absorption behaviour of cotton fibre-reinforced geopolymer composites

Table 5.1 Maximum water uptake and diffusion coefficient (D) of CF/geopolymer composites

Sample	CF content (wt%)	M_∞ (%)	$D \times 10^{-6}$ (mm^2/s)
Geopolymer (GP)	0	4.72	4.26
CF/GP1	4.5	11.74	5.42
CF/GP2	6.2	17.98	6.16
CF/GP3	8.3	22.32	8.4

active. The capillary mechanism involves the flow of water molecules along fibre–matrix interfaces and diffusion through the bulk matrix. Water molecules actively attack the interface, resulting in de-bonding of the fibre and the matrix (Dhakal et al. 2007).

The effect of water absorption on the mechanical properties of cotton fabric-reinforced geopolymer composites was investigated after placing specimens in water for 133 days at room temperature and comparing them with samples of the same composites kept in dry conditions.

5.1.2 Flexural Strength

The effect of fibre content on the flexural strength of dry CF reinforced geopolymer composites is shown in Fig. 5.2. In dry condition, the flexural strength increased as fibre content increased. The flexural strength of neat geopolymer increased from 8.3 to 15.8, 19.7 and 28.1 MPa due to the addition of 4.5, 6.2 and 8.3 wt% CF, respectively. This enhancement in flexural strength of CF-reinforced geopolymer composites is due to the ability of natural fibre to resist bending forces and good stress transfer from the matrix resulting in improve strength properties (Reis 2006a, b).

The effect of water absorption on flexural strength of CF reinforced geopolymer composites is also shown in Fig. 5.2. The flexural strength of composites decreased

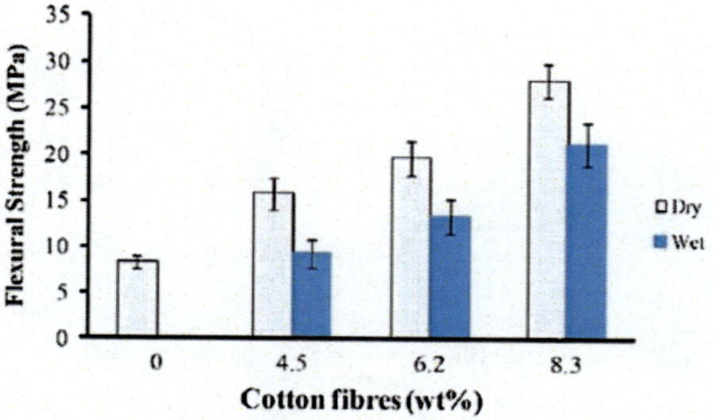

Fig. 5.2 Flexural strength of geopolymer composites in dry and wet conditions

markedly after water absorption. Compared to the dry composites, the flexural strength of the composites reinforced with 4.5, 6.2 and 8.3 wt% CF deceased from 15.8, 19.7 and 28.1 MPa to 9.3, 13.4 and 21.4 MPa, respectively. This could be because the immersion of the composite samples in water affects the interfacial adhesion between fibre and matrix and creates de-bonding, leading to a decrease in mechanical properties. When the fibre–matrix interface was accessible to moisture in the environment the cotton fibres swelled. This resulted in the development of shear stress at the interface and led to the ultimate de-bonding of the fibres, delamination and loss of structural integrity (Ghosh et al. 2011).

Water absorbed in polymers is generally as either free water or bound water as reported by Azwa et al. (2013) (see Fig. 5.3). Water molecules which are relatively free to travel through the micro voids and pores are identified as free water, while those dispersed in the polymer matrix and attached to the polar groups of the polymer are designated as bound water (Azwa et al. 2013). In a wet environment, water molecules penetrate in natural fibre-reinforced composite through micro-cracks and reduce interfacial adhesion of fibre with the matrix. This causes swelling of the fibres, which may create micro-cracks in the matrix and eventually lead to debonding between the fibre and the matrix (Girisha et al. 2012). A schematic illustration of this process is presented in Fig. 5.4.

Dry cotton fibre constructed from fibrils of cellulose is stiff and rigid. The cellulose molecules are held tightly together inside the fibrils by bonds established between

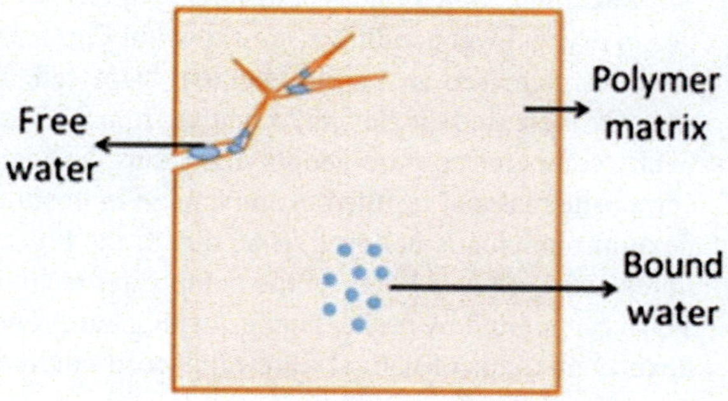

Fig. 5.3 Free water and bound water in polymer matrix (Adapted from Azwa et al. 2013)

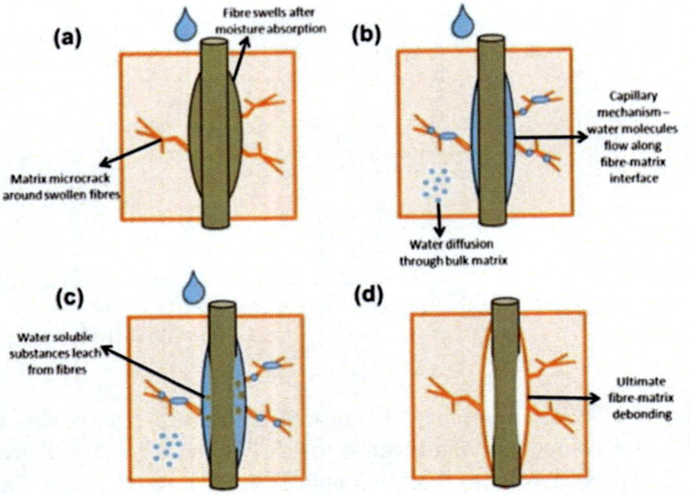

Fig. 5.4 Effect of water on fibre-matrix interface (Adapted from Azwa et al. 2013)

molecules lying closely alongside one another. Water, however, can penetrate this cellulose network and move into the capillaries and spaces between the fibrils. In this situation, water molecules tend to force the cellulose molecules apart, reducing the forces that hold them together and destroying their rigidity, because water acts as a plasticiser and permits the cellulose molecules to move. Consequently, the mass of the cellulose is softened, and this changes the dimensions of the fibre under applied force (Dhakal et al. 2007; Athijayamani et al. 2009). According to Ray and Rout (2005), water molecules attract the hydrophilic groups of natural fibres and react with the hydroxyl groups (–OH) of the cellulose molecules to form hydrogen bonds. A schematic illustration of moisture absorption by natural fibres is presented in Fig. 5.5.

5.1.3 Flexural Modulus

The flexural modulus values of different cotton fibre reinforced geopolymer composites in dry and wet conditions are shown in Fig. 5.6. In the dry samples, the flexural modulus increased as the fibre content increased. The addition of 4.5, 6.2 and 8.3 wt% CF increased the flexural modulus from 0.87 to 1.23, 1.4 and 1.74 GPa, respectively, compared to pure geopolymer: thus, an increase in the fibre content of the composite material resulted in an increase in flexural modulus. The improvement in flexural modulus is believed to be due to the higher initial modulus of the natural fibres acting as backbones in the composites (Kim and Seo 2006; Lee et al. 2009). This is supported by earlier studies, which have reported significant increases in the flexural modulus of natural fibre-reinforced polymer composites. For example, Ma

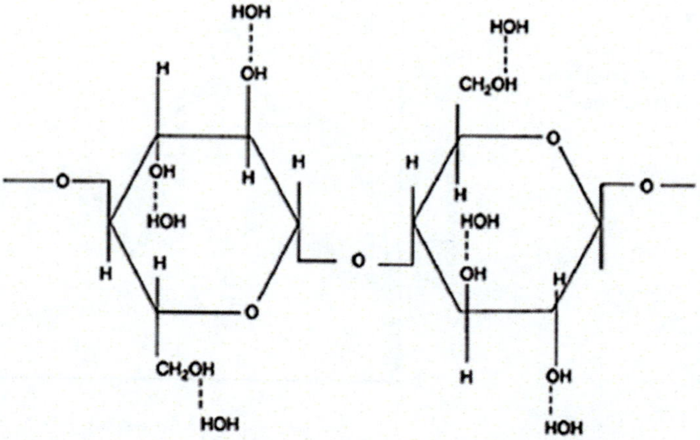

Fig. 5.5 Schematic of moisture absorption by natural fibre (Adapted from Ray and Rout 2005)

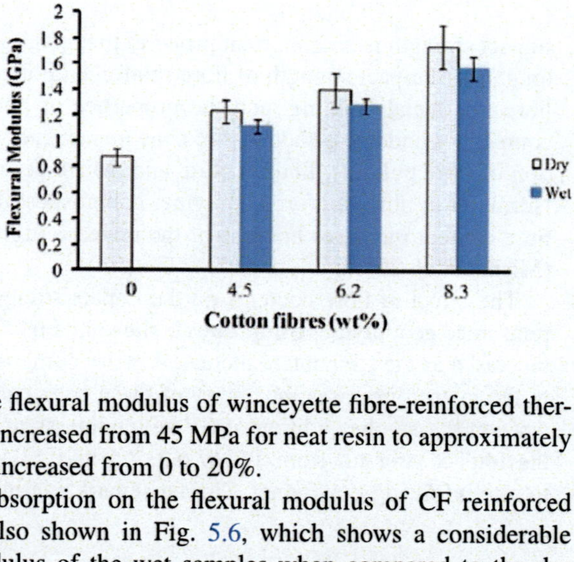

Fig. 5.6 Flexural modulus of geopolymer composites in dry and wet conditions

et al. (2005) reported that the flexural modulus of winceyette fibre-reinforced thermoplastic starch composites increased from 45 MPa for neat resin to approximately 140 GPa as the fibre content increased from 0 to 20%.

The influence of water absorption on the flexural modulus of CF reinforced geopolymer composites is also shown in Fig. 5.6, which shows a considerable decrease in the flexural modulus of the wet samples when compared to the dry samples. The reason for this is that in the wet samples absorbed water molecules and reduced the intermolecular hydrogen bonding between cellulose molecules in the fibre and established intermolecular hydrogen bonding between the cellulose molecules and water molecules in the fibre, thereby reduced the interfacial adhesion between the fibre and the matrix and resulting in decreased flexural modulus (Dhakal et al. 2007). Figure 5.7 illustrates the typical flexural stress–strain curves for geopolymer composites before and after being placed in water. It can be observed that the maximum stress in dry composite significantly decreased after immersion in water for a prolonged period. This drop can be attributed to degradation in the fibre–matrix interfacial bonding caused by the water absorption (Alamri et al. 2012).

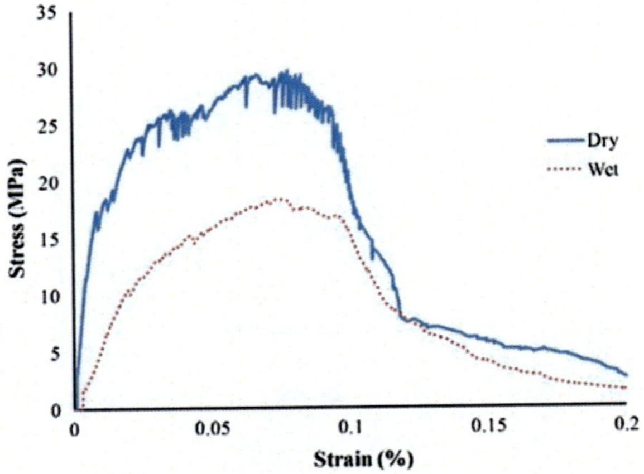

Fig. 5.7 Typical stress–strain curves of geopolymer composites in dry and wet conditions

5.1.4 Impact Strength

Impact strength is an important property that gives an indication of overall material toughness. Impact strength of fibre-reinforced polymer is governed by the matrix–fibre interfacial bonding, and the properties of both matrix and fibre. When the composites undergo a sudden force, the impact energy is dissipated by the combination of fibre pullouts, fibre fracture and matrix deformation (Wambua et al. 2003). Normally in fibre-reinforced polymer composites, the impact strength increases as fibre content increases because of the increase in fibre pull out and fibre breakage (Mishra et al. 2003).

The effect of fibre contents on the impact strength of dry and wet cotton fibre-reinforced geopolymer composites is shown in Fig. 5.8. Impact strength significantly increased as the CF content increased in dry composites. The presence of CF layers in the matrix increases the ability of these composites to absorb impact energy. In dry conditions, the addition of CF with contents of 4.5, 6.2 and 8.3 wt% increases the impact strength from 1.9 to 6.2, 8.5 and 13.4 kJ/m^2, respectively compared to unreinforced geopolymer. Similar remarkable improvements in impact strength

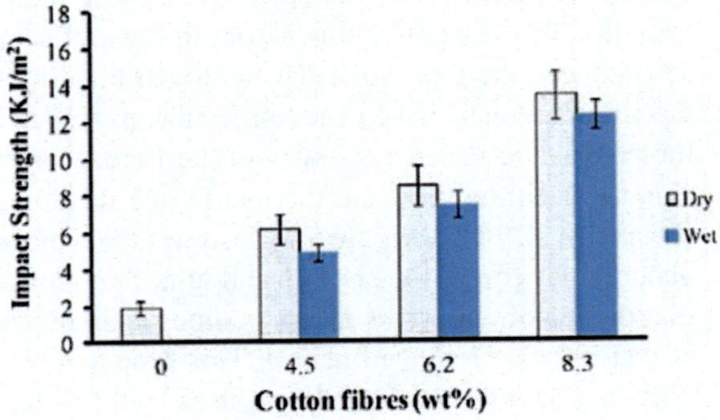

Fig. 5.8 Impact strength of geopolymer composites in dry and wet conditions

were reported by Graupner (2008), who observed that the addition of cotton fibre increased the impact strength of pure poly (lactic acid) (PLA) matrix. He concluded that the increase was due to greater elongation of cotton fibres at break. Fibres containing much cellulose generally have high elongation at break values. Cotton has a cellulose content of about 88–96%. Elongation at break and impact strength are directly correlated. The high elongation at break of cotton fibres increased the elongation at break in the composites, leading to higher impact strength.

However, impact strength is adversely affected by water absorption. The decrease in impact properties after water immersion can be related to the weak fibre–matrix interface, which resulted in a reduction of the mechanical properties and dimensional stability of composites (Dhakal et al. 2007).

5.1.5 Hardness

The effect of cotton fibre contents on the hardness of the cotton fibre-reinforced geopolymer composites is presented in Fig. 5.9. The hardness of geopolymer composites reinforced with 4.5, 6.2 and 8.3 wt% CF increased from 65.5 to 87.22, 92.32 and 86.4 HRH, respectively relative to the neat geopolymer. This enhancement in hardness is caused by the distribution of the test load on the fibres, which decreased the penetration of the test ball on the surface of the composite material and consequently improved the hardness of this material (Al-Mosawi 2009).

However, hardness is affected by water absorption, as shown in Fig. 5.9. Hardness decreases in all cotton fibre-reinforced samples in wet condition and is associated with the weakening of interface between the geopolymer matrix and the cotton fibre caused by the water absorption. This decrease has been reported by other researchers working with natural fibre-based composites. Dhakal et al. (2013) reported that as water absorption increased, the hardness of flax fibre-reinforced composites decreased, and found that the deformation depth increased for water-immersed specimens compared to dry ones, due to the hydrophilic nature of the fibres, and eventually led to the formation of a weak fibre–matrix interface. In the case of cotton fibre-reinforced

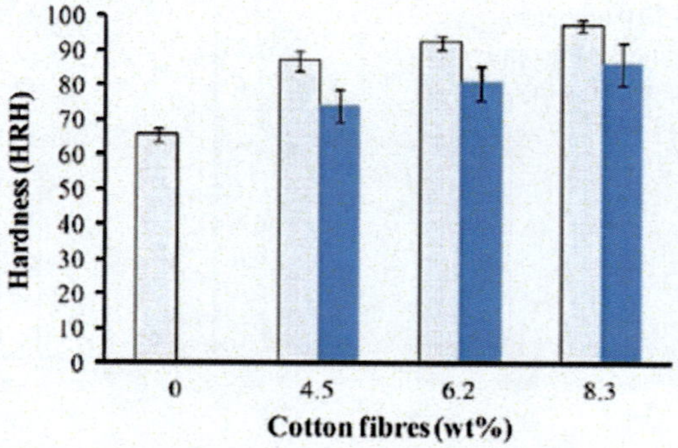

Fig. 5.9 Hardness of geopolymer composites in dry and wet conditions

geopolymer composites when water uptake reaches saturation level, the bound water and the free water remain in the composite as a reservoir. This leads to softening of the fibres and weakening of the fibre matrix adhesion, resulting in reduced material properties.

5.1.6 Fracture Toughness

The effect of cotton fibre contents on the facture toughness of geopolymer composites is presented in Fig. 5.10. The addition of cotton fibre gradually increased the fracture toughness of CF reinforced geopolymer composites compared to net geopolymer. Cotton fibres play a significant role in enhancing the facture toughness of the matrices through several energy-absorbing characteristics such as fibre rupture, fibre–matrix interface debonding, fibre pull-out and fibre-bridging, which slow crack propagation and therefore increase the fracture energy (Toledo Filho et al. 2000; Filho et al. 2003; Reis 2006a, b; de Andrade Silva et al. 2009; Silva et al. 2010). The fracture toughness of geopolymer reinforced with 4.5, 6.2 and 8.3 wt% CF increased from 0.57 to 1.09, 1.27 and 1.58 MPa $m^{1/2}$, respectively compared to neat geopolymer. This significant enhancement in facture toughness at higher CF content is due to extensive fibre pull-out, fibre fracture and fibre-bridging of cotton fibres.

The effect of water absorption on fracture toughness of CF-reinforced geopolymer composites is also shown in Fig. 5.10. The fracture toughness for all wet composites considerably decreased compared to the dry composites, as a result of the severe damage to fibre structure and interfacial bonding between the cotton fibre and the geopolymer matrix caused by the absorbed water.

Typical load–displacement curves for the composites before and after immersion in water are shown in Fig. 5.11. The maximum peak load of dry composite significantly decreased after immersing in water for a prolonged period. The areas under the curve indicate that the wet composite achieved lower fracture toughness than the dried composite. This reduction can be explained as the effect of moisture absorption causing swelling of fibres, which creates micro-cracks in the sample, leading to

Fig. 5.10 Fracture toughness of geopolymer composites in dry and wet conditions

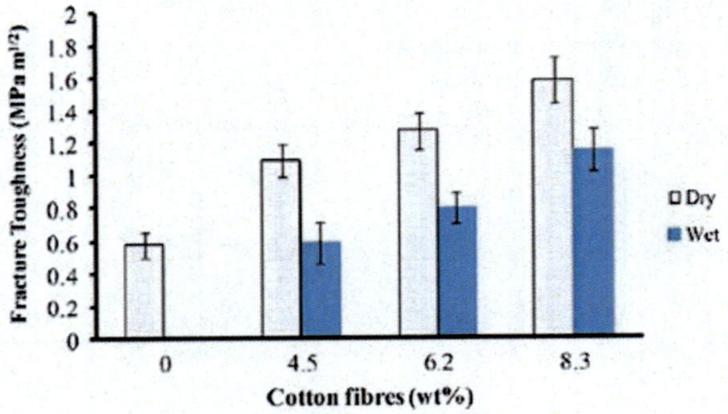

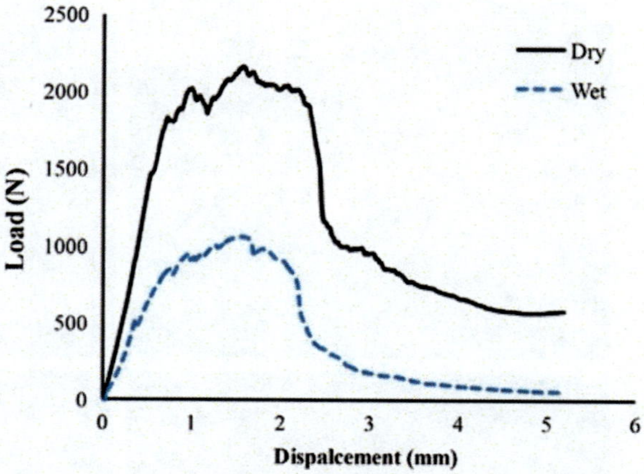

Fig. 5.11 Typical load–displacement curves of geopolymer composites in dry and wet conditions

lower fracture toughness (Reis 2006a, b). In addition, water molecules diffuse into the fibre–matrix interfaces through these micro-cracks, which cause debonding of the fibres and thus weakens the fibre–matrix interface (Dhakal et al. 2007).

The reduction in fracture toughness can be attributed to internal pore water pressure which developed in the limited pore spaces of the wet geopolymer composites. Water does not move into a pore when adjacent pores are filled with water. As a result, a very high disjoining pressure is produced due to the capillary action, leading to early crack propagation under external loading. The fracture resistance of wet geopolymer composites thus becomes lower than that of dry composites (Lau and Büyüköztürk 2010; Ren et al. 2017).

The microstructures of dry and wet composites reinforced with 8.3 wt% CF are shown in images of Fig. 5.12a–d. Images in Fig. 5.12a–b show severe matrix cracking and degradation of the interfacial adhesion between the fibres and the matrix in wet composites characterized by the appearance of gap between fibre and matrix. Water penetrates the cotton fibre bundle and causes the breaking down of the composite fibre bundle into finer fibrils due to decrease in bundle coherence when subjected to flexural loads, as shown in Fig. 5.12a. It can also be observed extensive fibre pull-out and no evidence or traces of matrix adhering to the fibre which are an indication of poor fibre-matrix adhesion as shown in Fig. 5.12c of wet composite. In contrast, prior to exposure to water, SEM micrographs showed almost no fibre pull-out, undamaged fibre bundle and small pieces of geopolymer paste were attached to the fibre surface of cotton fibre. These observations are indicative of strong bond between the fibres and the matrix in dry composite as shown in Fig. 5.12d.

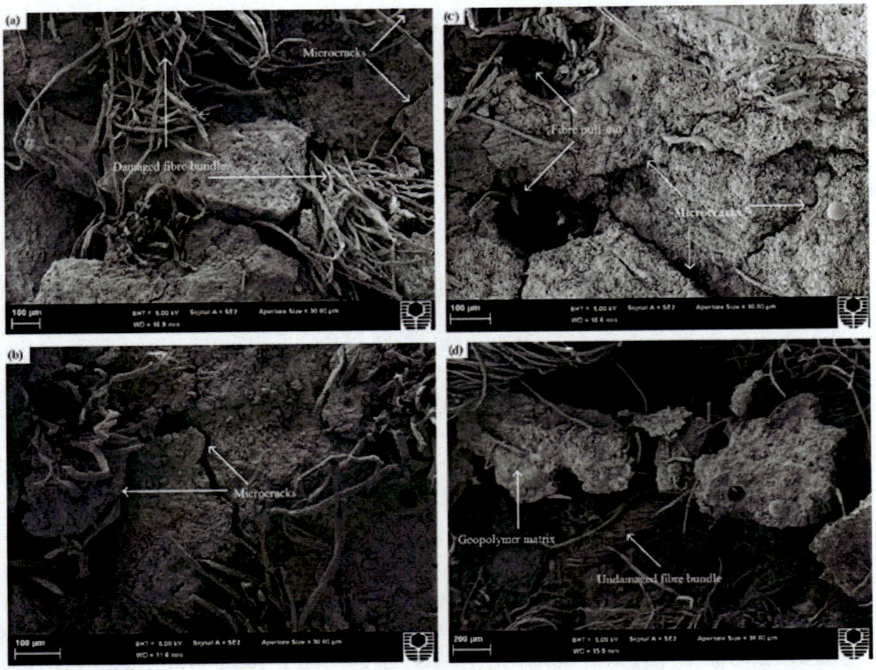

Fig. 5.12 SEM micrographs showing **a** separation of the cotton fibre bundles into finer fibrils, **b** matrix cracking, **d** fibre pull-out and **c** small pieces of matrix attached to the fibre

5.2 Flax Fabric-Reinforced Geopolymer Composites

5.2.1 *Effect of Nanoclay*

Density and Porosity

Table 5.2 shows the compositions of the fabricated geopolymer paste and geopolymer nanocomposites and their values of porosity and water absorption are shown in Table 5.3. Geopolymer nanocomposites revealed denser matrices and lower porosities when compared to the control sample. The addition of NC has increased the density and reduced the porosity of geopolymer nanocomposites when compared to control geopolymer paste. The optimum addition was found as 2.0 wt% of NC, which increased density by 11.4% and reduced the porosity by 7.2% when compared to the control paste. This implies that the nanoparticles played a pore-filling role to reduce the porosity of geopolymer composites. However, adding excessive amounts of NC increased the porosity and decreased the density of all samples due to agglomeration of NC particles (Assaedi et al. 2017). This finding is comparable with the study where the porosity of cement paste is decreased due to addition of 1.0% wt of NC to

Table 5.2 Formulation of samples

Sample	NC (g)	FF (layers)
GP	0	0
GPNC-1	10	0
GPNC-2	20	0
GPNC-3	30	0
GP/FF	0	10
GPNC-1/FF	10	10
GPNC-2/FF	20	10
GPNC-3/FF	30	10

Each samples is a mix of: 1.0 kg Eraring Flyash, 214.5 g sodium hydroxide (8 M) and 535.5 g sodium silicateEach samples is a mix of: 1.0 kg Eraring Flyash, 214.5 g sodium hydroxide (8 M) and 535.5 g sodium silicate

Table 5.3 Density and porosity for pure geopolymer and geopolymer nano-composites

Sample	Density (g/cm^3)	Porosity (%)
GP	1.84 ± 0.02	22.2 ± 0.4
GPNC-1	1.92 ± 0.02	21.3 ± 0.3
GPNC-2	2.05 ± 0.02	20.6 ± 0.3
GPNC-3	1.98 ± 0.03	21.0 ± 0.2

cement paste; however, the porosity is increased because of the agglomeration effect when more nanoparticles were added (Hakamy et al. 2015).

X-Ray Fluorescence (XRF) and X-Ray Diffraction (XRD)

The chemical composition and loss on ignition of flyash and NC are shown in Table 5.4. Flyash and NC contain, in addition to silica and alumina, Fe_2O_3, CaO, K_2O, Na_2O, MgO and TiO_2. The XRD spectra of pure geopolymer and geopolymer nanocomposites at 4 and 32 weeks are shown in Fig. 5.13a, b, respectively. The crystalline phases were indexed using Powder Diffraction Files (PDFs) from the Inorganic Crystal Structure Database (ICSD). The diffraction patterns of the samples demonstrate some crystalline phases that were indexed distinctly: quartz [SiO_2] (PDF 00-046-1045) and mullite [$Al_{2.32}Si_{0.68}O_{4.84}$] (PDF 04-016-1588). Quartz and mullite crystalline phases can be seen in all samples. According to Rickard et al. quartz and mullite are the main crystalline content of the Eraring flyash, and hence they are stable and unreactive in the alkaline environment. At 32 weeks, a new crystalline

Table 5.4 Chemical compositions of flyash and nanoclay (wt%)

	SiO_2	Al_2O_3	CaO	Fe_2O_3	K_2O	MgO	Na_2O	P_2O_5	SO_3	TiO_2	MnO	BaO	LOI
Flyash	63.13	24.88	2.58	3.07	2.01	0.61	0.71	0.17	0.18	0.96	0.05	0.07	1.45
NC	47.05	16.24	0.29	3.42	0.03	1.75	0.19	0.01	0.11	0.08	0.00	0.00	30.61

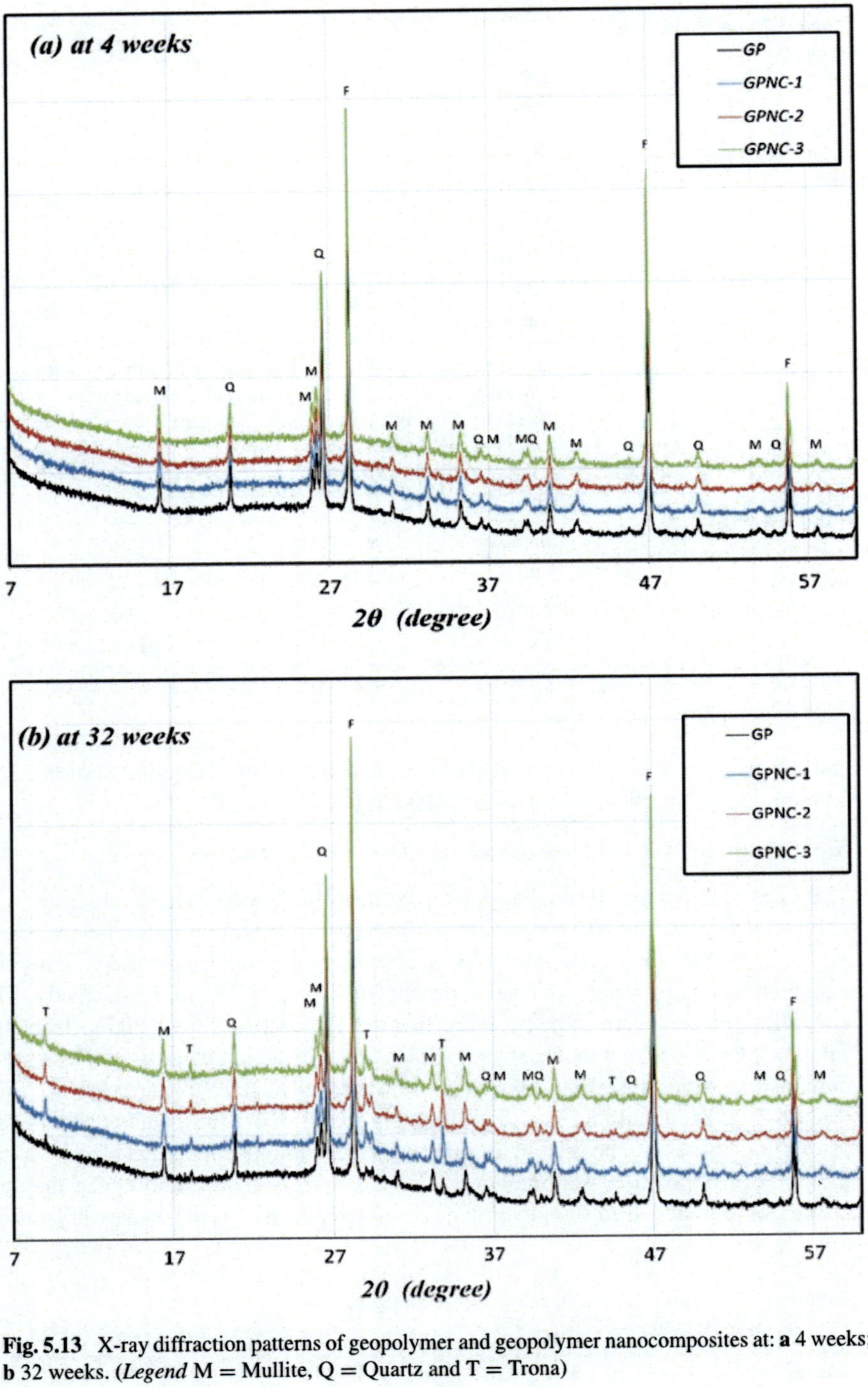

Fig. 5.13 X-ray diffraction patterns of geopolymer and geopolymer nanocomposites at: **a** 4 weeks; **b** 32 weeks. (*Legend* M = Mullite, Q = Quartz and T = Trona)

phase, trona [$Na_3H(Co_3)_2.2H_2O$] (PDF 00-029-1447), appears on the surface of geopolymer aged samples. Trona belongs to soda minerals group, which could be formed by the reaction of sodium hydroxide with water and carbon dioxide according to the chemical reaction (Hakamy et al. 2015):

$$3NaOH + 2CO_2 + 2H_2O \rightarrow Na_3H(CO_3)_2.2H_2O$$

The amorphous broad phase generated between $2\theta = 17°$ and 30° for all samples reveals the reactivity of geopolymers. It is known that the amorphicity degree remarkably influences the mechanical properties of geopolymers. When the amorphous content is higher, the strength of geopolymers is similarly higher (Graupner 2008). In previous study, it has been shown that the addition of nanoclay particles to geopolymer pastes increased the amorphous content of geopolymer nanocomposites, resulting in denser matrices and superior mechanical performance (Assaedi et al. 2017).

FTIR Observation

FTIR spectra of pure geopolymer and geopolymer nanocomposite at 4 and 32 weeks are shown in Fig. 5.14a, b. The FTIR spectra of all samples shows a strong peak at ~1000 cm^{-1} which is attributed to Si–O-Si and Al–O–Si asymmetric stretching vibrations, which is the identification peak of the geopolymerisation (Phair and Van Deventer 2001; Li et al. 2012). A broad peak in the region around 3400 cm^{-1} indicates that the OH group is present attached to different centers (Al, Si) and free water (Chindaprasirt et al. 2009; Rattanasak and Chindaprasirt 2009). The absorbance peak at 1640 cm^{-1} is also attributed to the (OH) bending vibration (Ul Haq et al. 2014). At 32 weeks changes have occurred, two peaks at 1420 and 1480 cm^{-1} appear indicating the presence of sodium carbonate; this was formed due to the atmospheric carbonation on the matrices surfaces which confirms the XRD results (Hakamy et al. 2015). During the aging period the reaction has carried on at a low rate consuming more OH groups and forming stronger material. The water content decreased to some equilibrium level during this period resulting in lower broad peak at 3400 cm^{-1}.

Flexural Strength of Geopolymer Nanocomposites

The effect of aging on the flexural strength and modulus of geopolymer matrices and nanocomposites is shown in Fig. 5.15. Overall, the incorporation of nanoclay into the geopolymer composite led to noteworthy improvement in the mechanical strength at all ages. At 4 weeks, the flexural strength of geopolymer nanocomposite containing 1.0, 2.0 and 3.0 wt% NC was increased by 13.3, 24.4 and 15.5% respectively, while the flexural modulus improved by 16, 25 and 20%, respectively compared to the control sample. This enhancement noticeably shows the value of NC in supporting geopolymer reaction and filling the micro pores in the matrix (Hakamy et al. 2015; Assaedi et al. 2017). Thus, the microstructure of geopolymer nanocomposite is denser than the pure matrix, especially in the case of incorporating 2.0 wt% NC, which is evident from its higher flexural strength and modulus. However, at 32 weeks, the flexural strength of nanocomposites increased slightly compared to their values at

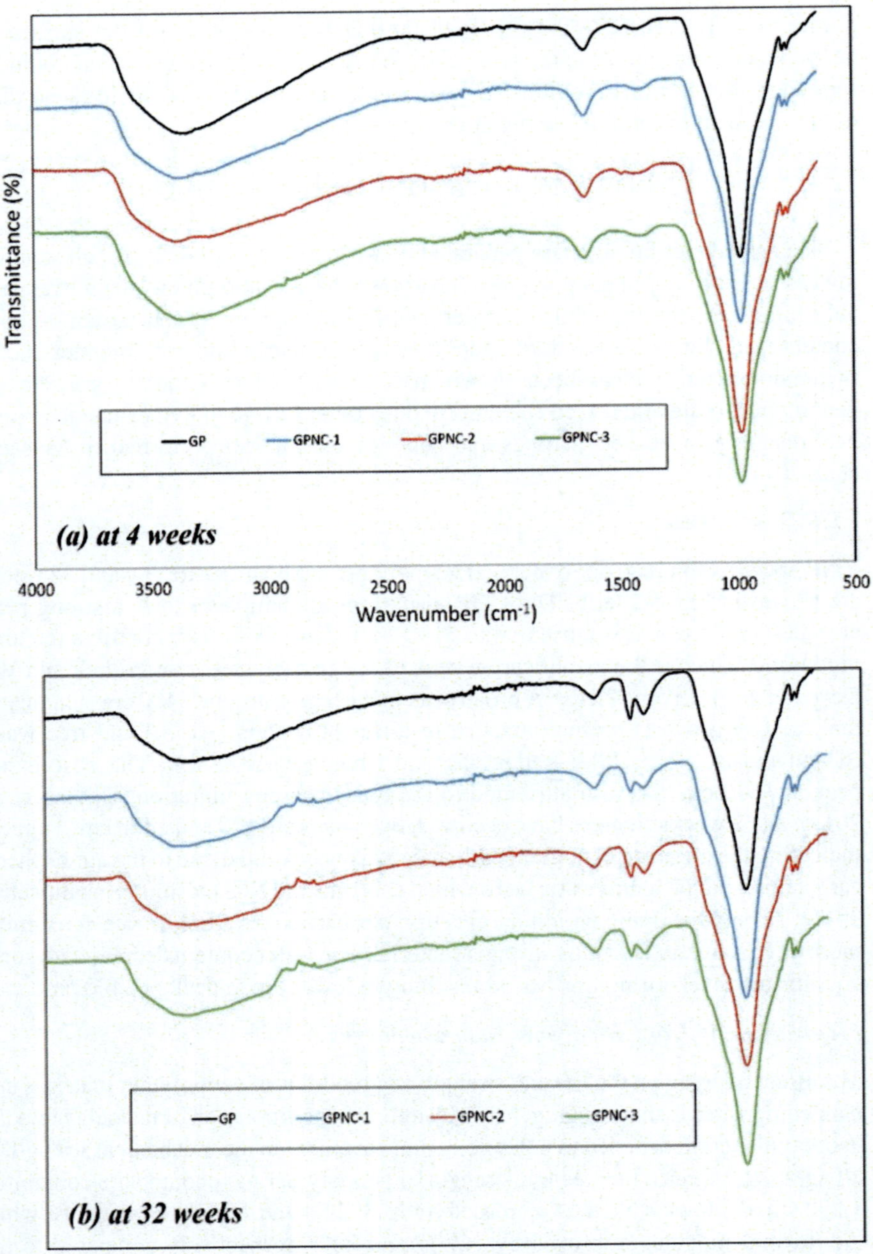

Fig. 5.14 FTIR spectra of geopolymer and geopolymer nanocomposites at: **a** 4 weeks; **b** 32 weeks

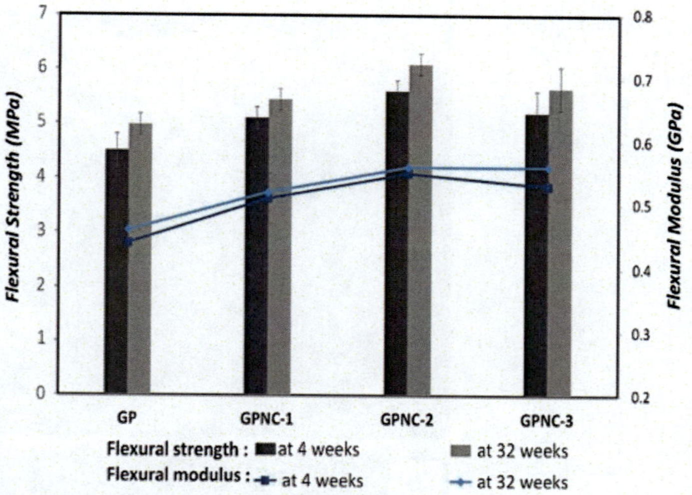

Fig. 5.15 Flexural strength and modulus of geopolymer and nanocomposites at 4 and 32 weeks

4 weeks. For instance, the flexural strength of GPNC-2 nanocomposite improved from 5.6 to 6.1 MPa by about 9% increase. This slight improvement in the mechanical performance could be attributed to the slow reaction of free silica and alumina in the presence of Na^{+} ions during the aging period (Mohr et al. 2006; Yan and Chouw 2015).

In similar study, Hakamy et al. (2016) reported that flexural strength of cement pastes containing 1.0% calcined nanoclay particles improved from 7.2 to 8.2 by about 7% after 236 days compared to its strength at 56 days. SEM images of the microstructure at 32 weeks of geopolymer paste and the geopolymer nanocomposite containing 2.0 wt% NC are shown in Fig. 5.16a–b. For geopolymer matrix, Fig. 4a displays more pores showing a weak microstructure. On the other hand, Fig. 4b shows the SEM micrograph of GPNC-2 nanocomposite matrix, which is different from that of pure matrix, the microstructure is denser and more compact with fewer pores and more geopolymer gel.

Flexural Strength of Flax Fabric Reinforced Geopolymer Nanocomposites

The effect of aging on the flexural strength and modulus of FF-reinforced geopolymer nanocomposites at 4 and 32 weeks is shown in Fig. 5.17. The incorporation of nanoclay into matrices led to enhancement in the flexural strength of all reinforced nanocomposites. For example, at 4 weeks, the flexural strength and modulus of GPNC-2/FF increased by 32.4 and 5.2%, respectively when compared to GP/FF composite. However, all composite showed reduction in the mechanical strength after 32 weeks. Plots in Fig. 5.18a, b show the effect of aging on the load-midspan deflection behaviour of GP/FF composites and GPNC-3/FF nanocomposites. The "ductile" behavior can be observed in both composites with and without NC, with higher load capacity (about 29% increases) in the composite containing NC. It was

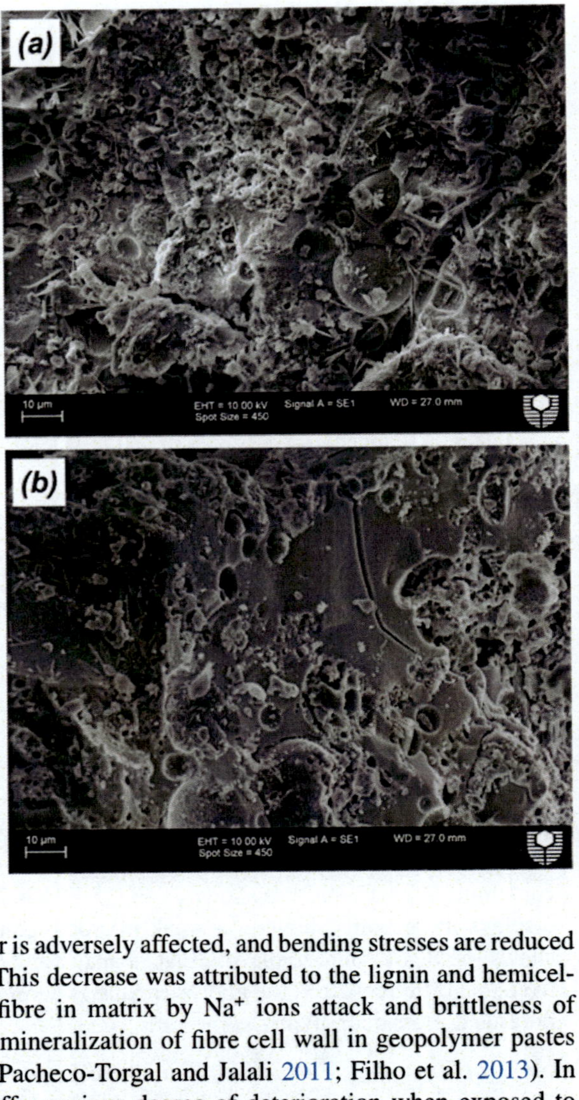

Fig. 5.16 SEM micrographs at 32 weeks of: **a** geopolymer paste, **b** nanocomposites containing 2.0 wt% NC

observed that ductile behavior is adversely affected, and bending stresses are reduced due to degradation process. This decrease was attributed to the lignin and hemicellulose deterioration of flax fibre in matrix by Na^+ ions attack and brittleness of the natural fibres due to the mineralization of fibre cell wall in geopolymer pastes (Bentur and Mindess 2007; Pacheco-Torgal and Jalali 2011; Filho et al. 2013). In general, all-natural fibres suffer various degree of deterioration when exposed to alkaline environment. This degradation in alkali matrices ultimately led to weaken the fibre–matrix bonding, and consequently reduced the mechanical performance of geopolymer composites. The flexural strength of GP/FF composite dropped to 23.0% of the initial strength at 4 weeks, whereas the flexural strength of GPNC-1/FF, GPNC-2/FF and GPNC-3/FF nanocomposites reduced by about 14.4, 13.7 and 13.5% compared to its value at 4 weeks. Based on this outcome, it can be concluded that the reduction in the mechanical performance for nanocomposites was less than the reduction of control sample composite after 32 weeks aging period. This may be attributed to the fact that nanoclay particles consume amounts of the alkaline

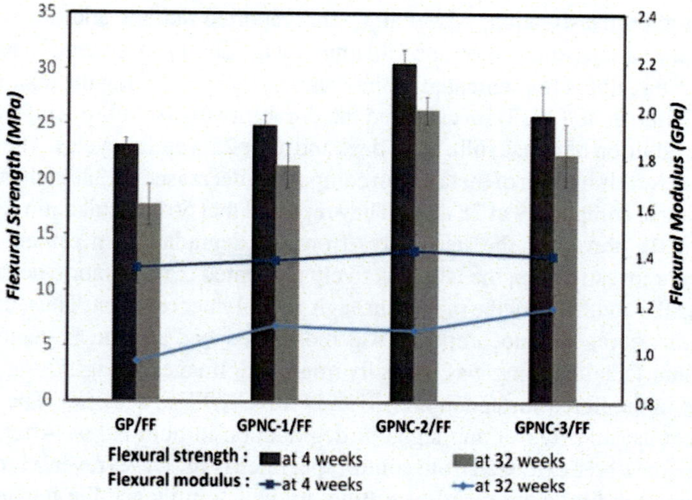

Fig. 5.17 Flexural strength and modulus of flax fabric reinforced geopolymer composites and nanocomposites at 4 and 32 weeks

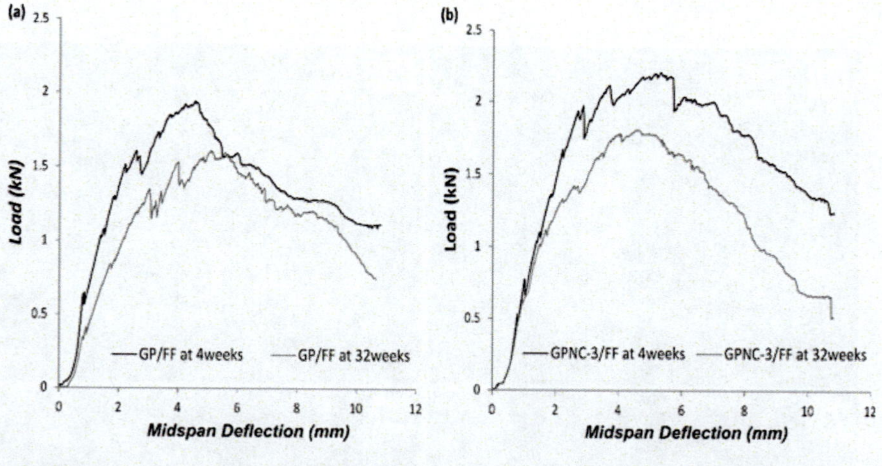

Fig. 5.18 Load versus mid-span deflection curves at 4 and 32 weeks for: **a** GP/FF composite; and **b** GPNC-3/FF nanocomposites

solution thus reducing the alkalinity of the medium, and producing higher amount of geopolymer gel, which hence enhances the density of the matrices and the fibre-matrix adhesion (Aly et al. 2011). In a similar study, the effect of calcined nanoclay on the durability of hemp fabric reinforced cement nanocomposites and the degradation of hemp fibres are reported (Hakamy et al. 2016), the nanoparticles were found to improve the durability and reduces the degradation of hemp fibres.

In another investigation, Aly et al. (2011) reported that the addition of nanoclay and waste glass to cement mortar could improve the durability and mitigate the degradation of flax fibres implemented in the composites by reducing the alkalinity of the matrix. Filho et al. (2013) investigated the durability of sisal fibre reinforced mortar with the addition of metakaolin at 28 days and after 25 wet/dry cycles. They observed that the flexural strength of metakaolin composites decreased by 23% when compared to its control composites at 28 days. They reported that 50% metakaolin replacement significantly prevented the sisal fibres from the degradation in cement matrix. In the current investigation, the NC effectively prevented the flax fabric degradation by reducing the alkalinity of the matrix through geopolymer reaction. Thus, the degradation of flax fibres in nanocomposite was mostly reduced and the FF-nanocomposite matrix interfacial bonding was typically improved. Images in Fig. 5.19a, b show the changes in the fibres surface in GP/FF and GPNC-2/FF at 4 weeks. The fibres look more regular, and free of any signs of degradation in both cases. After 32 weeks, however, the fibres extracted from control specimen (Fig. 5.19c) reveal signs of degradation and the fibrils are clearly splitting up, which influence the flexural strength of the composite, whereas the fibres extracted from GPNC-2/FF after aging period (Fig. 5.19d) do not present signs of significant damage. These desirable attributes are attractive for these materials to be used in civil engineering and building construction

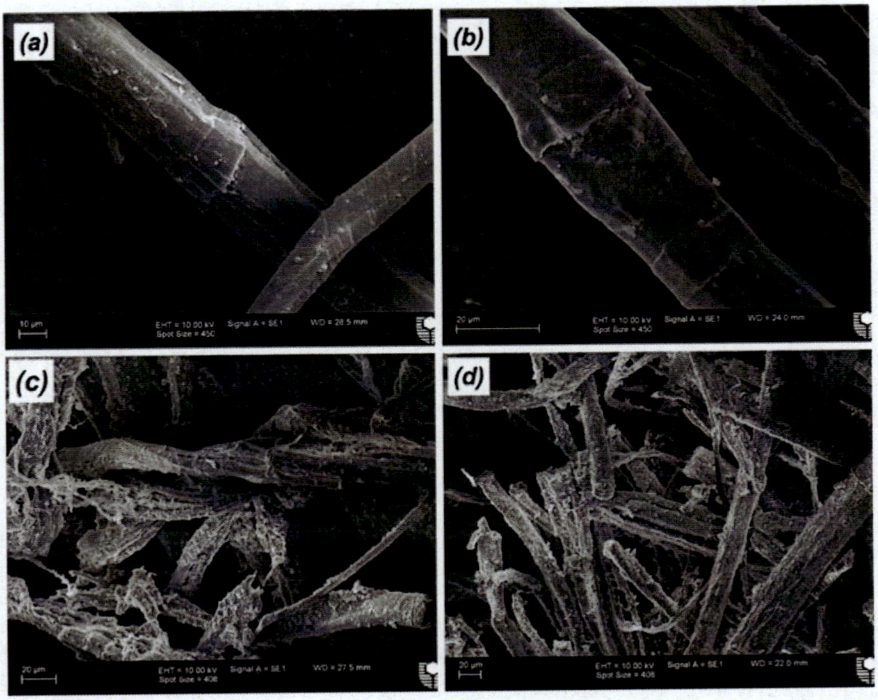

Fig. 5.19 SEM images of the flax fibres extracted from **a** GP/FF at 4 weeks, **b** GPNC-2/FF at 4 weeks, **c** GP/FF at 32 weeks, and **d** GPNC-2/FF at 32 weeks

industries. Possible applications include slabs or shingles for siding, certain types of roofing, and some interior uses in the building structure. They may also be used for other applications such as pipes and cooling towers.

5.2.2 *Effect of Nanosilica*

The fabricated samples are categorised in two series. The samples in first series was cured for four weeks under ambient conditions. The second series was stored under similar conditions but for 32 weeks. Table 5.5 illustrates mix proportions of the samples.

Physical Properties

The bulk density and porosity of pure geopolymer and that containing NS are shown in Table 5.6. The geopolymers containing NS exhibited higher density and lower porosity compared to the control sample. The ideal content was found as 1.0 wt% of nano silica, which creates denser matrix by about 15% and decreased the porosity by 27% as compared to the pure geopolymer. That improvement might be due to two

Table 5.5 Mix proportions of samples

Sample	Fly ash (g)	Alkaline Solution (g)	NS(g)	Water(g)	FF (wt%)
GP	1000	750	0	50	0
GPNS-0.5	1000	750	5	50	0
GPNS-1.0	1000	750	10	50	0
GPNS-2.0	1000	750	20	50	0
GPNS-3.0	1000	750	30	50	0
GP/FF	1000	750	0	50	4.1
GPNS-0.5/FF	1000	750	5	50	4.1
GPNS-1.0/FF	1000	750	10	50	4.1
GPNS-2.0/FF	1000	750	20	50	4.1
GPNS-3.0/FF	1000	750	30	50	4.1

Table 5.6 Density and porosity values of the control sample and geopolymer containing nano-silica

Sample	Density (gm/cm^3)	Porosity (%)
GP	1.84 (0.02)	22.20 (0.45)
GPNS-0.5	1.89 (0.02)	20.87 (1.35)
GPNS-1.0	2.10 (0.02)	16.08 (0.76)
GPNS-2.0	2.04 (0.03)	17.49 (1.84)
GPNS-3.0	1.96 (0.08)	20.33 (1.01)

Uncertainties are indicated in brackets

reasons. First, the nanoparticles filled the pores in the matrices which reduced the porosity of geopolymer nanocomposites. Secondly, the additional silica improved the geopolymeric reaction, thus creating further geopolymer gel and denser matrix accordingly (Phoo-ngernkham et al. 2014). However, further increase in NS content slightly reduced the density due to the inadequate dispersion and the agglomeration of the nano silica particles. The finding of the current study is consistent with that conducted by Supit and Shaikh (2014) who found that loading of 2.0 wt% NS considerably decreased the concrete porosity. In a similar investigation, the porosity of cement paste has been reduced due to addition of nano-clay particles to the samples; and the porosity increased with increase in nano-clay due to the influence of agglomeration (Hakamy et al. 2015). Images in Fig. 5.20a, b show the microstructures of pure geopolymer, and that contained 3.0 wt% NS. In the case of neat geopolymer sample, higher number of voids, un-reacted fly ash particles can be seen in Fig. 5.20a. However, less number of fly ash particles with higher amount of geopolymer gel appeared in the nanocomposite matrix (Fig. 5.20b).

Characterisation and Microstructure

The chemical compounds in Eraring fly ash as characterised by XRF were 63.1 wt% SiO_2, 24.8 wt% Al_2O_3, 2.58 wt% CaO and 3.07 wt% Fe_2O_3. The XRD spectra taken for the samples at four as well as 32 weeks are shown by the plots in Fig. 5.21a, b. Crystalline phases in the samples were identified in XRD analysis. Fluorite [CaF_2] was the standard used to calculate the weight fraction of each crystalline phase. At four weeks, all samples contained two crystalline phases, mullite [$Al_{2.32}Si_{0.68}O_{4.84}$] and quartz [$SiO_2$] (PDF 00-046-1045). Mullite and quartz can be observed in all specimens since they are the essential content in the Eraring fly ash. Thus they are both not reactive in the alkaline media (Assaedi et al. 2016b). However, a further crystalline content called trona [$Na_3H(Co_3)_2.2H_2O$] developed on the surface of the pure geopolymer and nanocomposites after 32 weeks. Trona is one of the soda

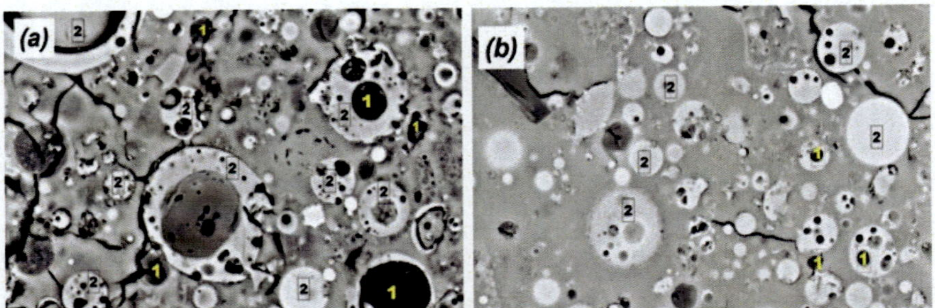

Fig. 5.20 SEM images showing the microstructure of **a** GP and **b** GPNS-3.0 (Assaedi et al. 2016a). Numbers indicate: 1 = voids; 2 = fly ash particles

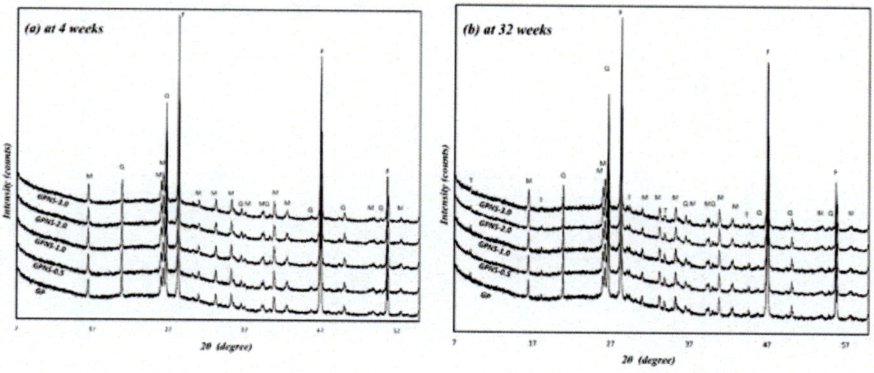

Fig. 5.21 X-ray diffraction patterns of the control sample and geopolymer with various contents of nano silica: **a** at four weeks; **b** at 32 weeks. Legend: F = Fluorite, M = Mullite, Q = Quartz and T = Trona

minerals group that formed by the atmospheric reaction of carbon dioxide with the residual water and sodium hydroxide in the system (Zaharaki et al. 2010).

The contents of crystalline phases compositions in the specimens were quantified by Rietveld analysis refinement procedure, and then the amorphous contents were computed. Plots in Fig. 5.22a, b show the refined XRD patterns of the neat geopolymer, and that contained 2.0% NS at 32 weeks, respectively. In general, the addition of amorphous nano silica to the geopolymer pastes led to minor changes in the crystalline and amorphous content in all specimens. Figure 5.23 presents the weighted percentage of crystalline (quartz, mullite and trona) and amorphous content in the neat matrix and geopolymer with NS at 4 and 32 weeks. At four weeks, the amorphous content of GPNS-3.0 has increased by 3.2 wt% mainly because of the additional 3.0 wt% NS, which caused a reduction in the crystalline phase's contents. Since nano silica is amorphous, the increase of the amorphous material in the nanocomposite specimens is typically because of the additional NS that was

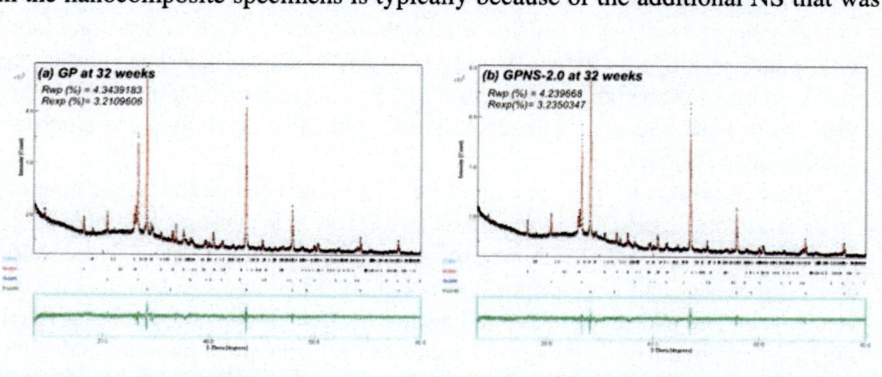

Fig. 5.22 XRD Rietveld refinements for **a** control sample and **b** GPNS-1 at 32 weeks. Experimental patterns are shown in black dots, and patterns as calculated in solid red lines. The green residual line displaying the variance between the patterns as calculated and measured experimentally

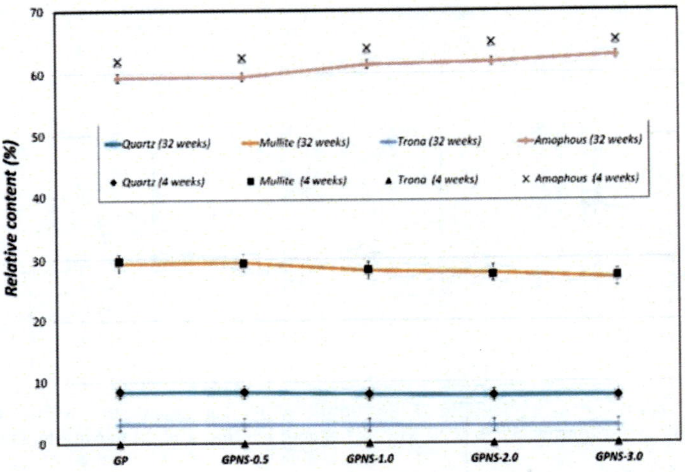

Fig. 5.23 Crystalline and amorphous phases structures of the neat geopolymer and that contained various contents of nano silica at 4 and 32 weeks

loaded to the pastes and performs as a nano-filler (Phoo-ngernkham et al. 2014; Nazari and Sanjayan 2015). The active and reacted nano silica particles could also promote the geopolymerisation reaction creating a higher amounts of amorphous geopolymeric gel within the matrices (Phoo-ngernkham et al. 2014). At 32 weeks, the amounts of quartz and mullite remained unchanged after the ageing period, those crystals are very stable and require high temperatures to be dissolved (Deer and Howie 1996; Schneider et al. 2008). Trona crystals formed mainly from the amorphous content in geopolymers and the atmospheric carbon dioxide. The formation of trona can be noticed by comparing the trona and amorphous content at 4 and 32 weeks. After the ageing period, the amorphous amount in geopolymers decreased by about 3% in all samples, which is similar to the amounts of trona that is grown after the ageing period. It was noted that the amount of carbonation content (trona) is highly dependent on the position is chosen from each sample since trona formed on the surfaces of geopolymers. However, all XRD samples have been chosen from similar parts in each composite to ensure they are nearly identical; hence, the analysis can provide a reasonable estimation of the relative crystalline and amorphous content in each sample.

Figure 5.24 shows FTIR spectra of the control sample and that containing nano silica at 4 and 32 weeks. The strong peak at 1000 cm^{-1} is interpreted to an overlap of Al–O–Si as well as Si–O-Si asymmetric stretching vibrations. This is widely known as a distinct peak and fingerprint of geopolymers (Phair and Van Deventer 2002; Li et al. 2012). The widespread peak in the range of about 3400 cm^{-1} can be ascribed to the hydroxy group (OH) linkage to various centres (mainly aluminium and silicon) and free water (Chindaprasirt et al. 2009; Rattanasak and Chindaprasirt 2009). The peak at 1640 cm^{-1} is an absorbent and likewise referred to the bending vibration of OH (Rattanasak and Chindaprasirt 2009). However, changes took place at 32 weeks,

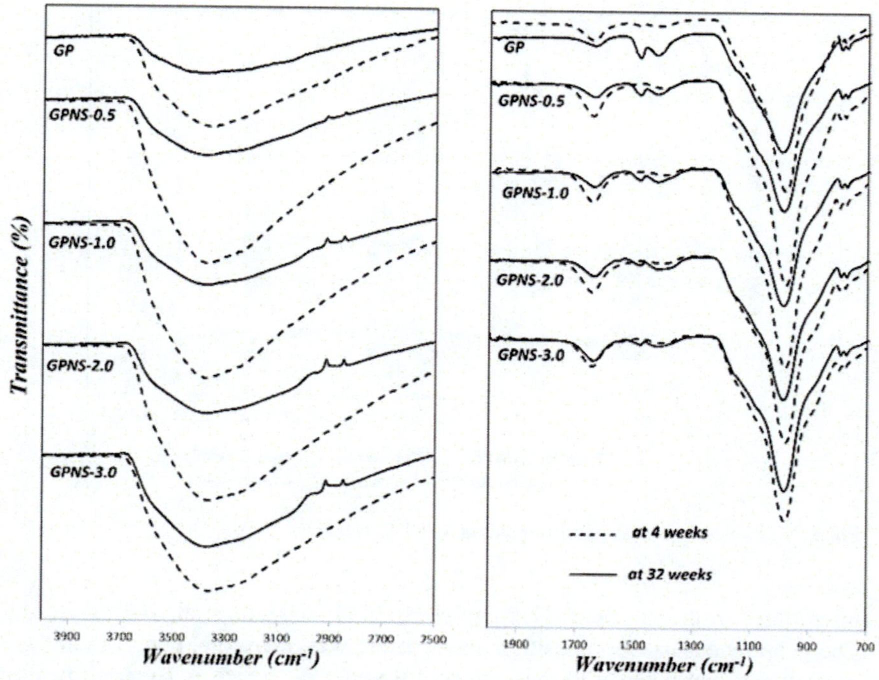

Fig. 5.24 Spectra of FTIR for geopolymer sample and that containing various contents of nano silica at 4 and 32 weeks

the development of sodium carbonate can be identified from the new two peaks at 1420 and 1480 cm^{-1}. Those peaks were created by atmospheric carbonation on the surface of the matrices, which confirms the positive results of the XRD (Zaharaki et al. 2010). During the ageing process, the geopolymeric reaction was carried on at a slow rate which consumed further hydroxy groups and formed tougher geopolymer. Throughout that period, the water content decreased slightly to a certain level of balance, resulting in a lower wide-ranging peak at 3400 cm^{-1}. Also, the geopolymer band at about 1000 cm^{-1} shifted slightly to higher wavenumbers due to a higher polycondensation reaction, thus indicating the higher transformation of Gel 1 to Gel 2 (Duxson et al. 2007; Liew et al. 2016).

Flexural Properties of Geopolymer Nanocomposites

Figure 5.25 illustrates how flexural strength (of nanocomposites as well as pure geopolymer) is impacted by ageing. Generally, a slight improvement in flexural strength was noticed at all ages due to the incorporation of nano silica particles into the geopolymer composite. In comparison with the control paste, the geopolymer nanocomposites in which the nano silica content was 0.5, 1.0, 2.0 and 3.0 wt% showed an increase in flexural strength (at four weeks) by approximately 20, 28, 24 and 15%, respectively. The improvement observed illustrates that nano silica particles are effective in enhancing geopolymeric reaction as well as in filling the

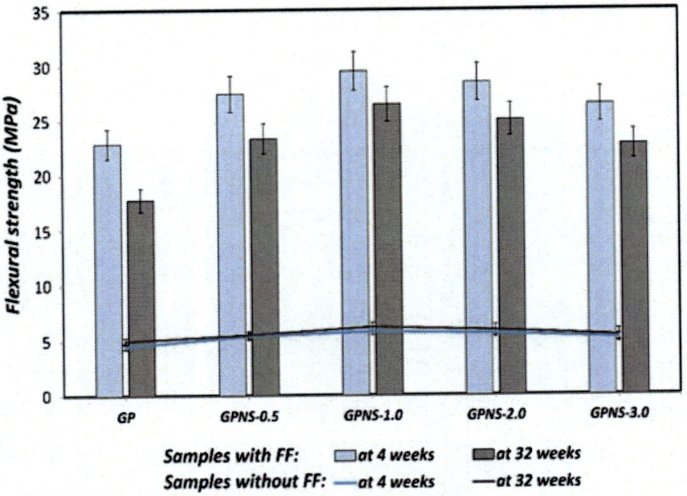

Fig. 5.25 Flexural strength of all samples at 4 and 32 weeks

micropores in the matrices (Hakamy et al. 2015; Assaedi et al. 2016b, c, 2019). The geopolymer nanocomposite microstructure was, therefore, denser than the net geopolymer, particularly in the case of 1.0 wt% NS, which is apparent from the result of the flexural strength. Once again, the flexural strength of nanocomposites increased slightly at 32 weeks compared to their early results at four weeks. For example, GPNS-1 nanocomposite's average flexural strength is improved from 5.8 to 6.2 MPa (6.8% increase). This small but noticeable improvement in mechanical properties can be related to the reaction of both alumina and silica during the ageing period in the alkaline environment (Zhang et al. 2010; Yadollahi et al. 2015). In a comparable investigation, Rong et al. (2015) studied the effects of the addition of 3.0 wt% of nano silica on the durability of concrete that contained 35 wt% fly ash in 28 and 90 days. They reported an increase in flexural strength of almost 11% at 90 days as opposed to 28 days. Another study (Mohamed 2016) also reported that after 90 days, the flexural strength of concrete, including 1.0 w.t% nano silica, increased by about 10% compared to 28 days.

Flexural Properties of Flax Fabric Reinforced Geopolymer Nanocomposites

The flexural strength investigations are repeatedly used to evaluate the mechanical performance of multilayered composites because they give a simple way of measuring the bending behaviour, providing technical results on the overall behaviour of the composites under applied force (Abanilla et al. 2006). Figure 5.25 shows the effect of the ageing process on the flexural strength of the samples at 4 and 32 weeks. The incorporation of nano silica particles into matrices has resulted in a significant increase in the flexural strength of all FF reinforced nanocomposites. For example, GPNS-1/FF's flexural strength at four weeks is increased from 23.0 to 30.5 MPa, approximately 32.4% higher than GP/FF samples. This improvement could be attributed to the enhanced density of the nanocomposites that led to better adhesion bond between flax fibres and geopolymer nanocomposites. The fracture surfaces of GP/FF and GPNS-0.1/FF after the flexural test at four weeks are shown in the images of Fig. 5.26a–c, respectively. The fracture surface of GP/FF composite shows high porosity, voids and particles of fly ash that are deeply rooted in the matrices. Thie increased porosity reduced the fibre-matrix bond, allowing the fibres to debound and pull out of the geopolymer matrix as seen in Fig. 5.28a. However, the fabric-reinforced nanocomposites loaded with 1.0 wt% nano silica exhibits relatively denser microstructure with less unreacted particles of fly ash. Consequently, better adhesion between fibres and the matrx is observed, causing the fibres to be fractured (Fig. 5.28b, c). In both cases, the fibres seemed uniform and did not show any visible signs of degradation.

After the ageing period of 32 weeks, all composites showed a reduction in the flexural test. Plots in Fig. 5.27a, b show the load–deflection behaviour of GP/FF and GPNS-1/FF composites at 4 and 32 weeks. The ductile behaviour can be seen with and without the nanoparticles in both composites, with higher load capacity (31%) in the composite containing NS. It was noticed that the degradation process causes the bending stress as well as ductile behaviour to decline. Normally cellulose fibres suffer

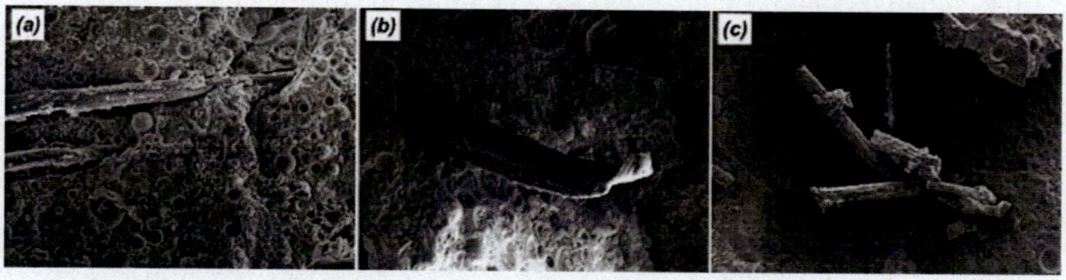

Fig. 5.26 SEM images showing the fracture surfaces of flax fibres reinforced composites; **a** GP/FF, and **b, c** GPNS-1/FF (Assaedi et al. 2016a)

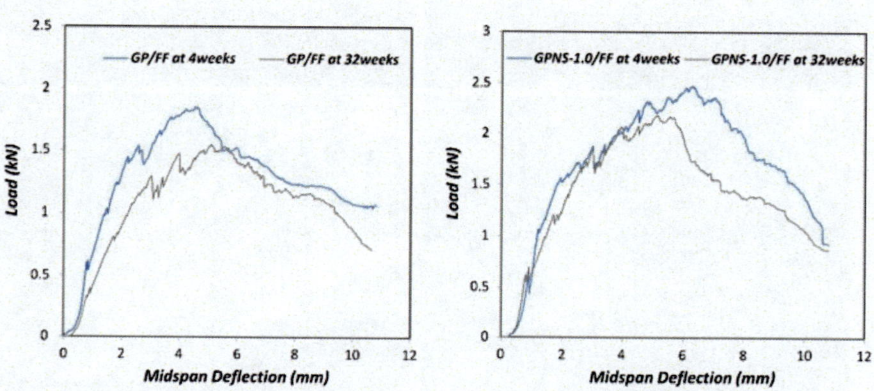

Fig. 5.27 Load–midspan deflection graph for GP/FF composite and GPNS-1/FF nanocomposites at 4 and 32 weeks

different levels of deterioration when exposed to the alkaline environment (Bentur and Mindess 2007). The degradation of natural fibres has been addressed by several studies and was attributed to the lignin and hemicellulose weakening due to the attack of alkali ions, and the mineralisation of fibre cell walls in geopolymer pastes that led to the fibre brittleness (Mohr et al. 2007; Pacheco-Torgal and Jalali 2011; Filho et al. 2013). The fibre degradation in alkali pastes eventually led to the deterioration of adhesion force that bound fibres to the matrices, and the concomitant reduction in flexural strength of the composites. The flexural strength of geopolymer composite reinforced with flax fibres reduced by 22.4% of its strength at four weeks, while the flexural strength of GPNS-0.5/FF, GPNS-1/FF, GPNS-2/FF and GPNS-3/FF samples is reduced by approximately 14.9, 10.3, 12.1 and 13.8%, respectively compared to the results at four weeks. From this results, it can be seen that after 32 weeks (of ageing), the nanocomposites experienced less reduction in flexural strength compared to the control sample composite. That could be because some amount of the alkaline solution was consumed by nano silica; therefore the alkalinity of the matrix is reduced, and the additional silica activated the geopolymerisation increasing the amount of geopolymer gelin the matrix. These phenomenon noticeably improved the matrix density and therefore the adhesion between the fibres and geopolymer matrix [26]. Images in Fig. 5.28a–d show the changes in the surface of the fibre in GP/FF, GPNS-1.0/FF and GPNS-3.0/FF at 32 weeks. The fibres in control sample (Fig. 5.28a, b) revealed degradation signs and separation of the small fibrils, which influenced the

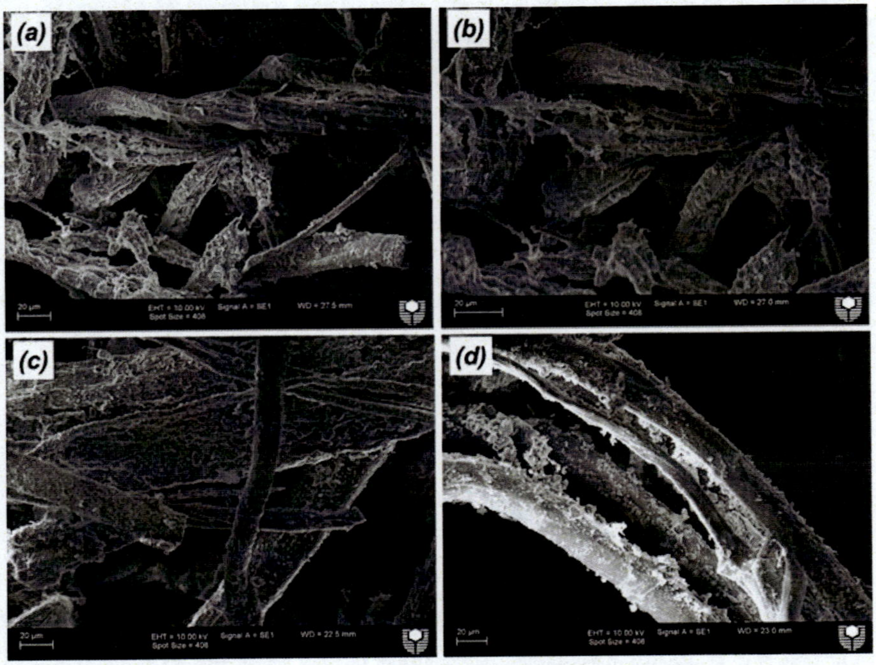

Fig. 5.28 SEM images of the flax fibres at 32 weeks in: **a**, **b** GP/FF, **c** GPNS-1/FF and **d** GPNS-3/FF

flexural behaviour of the composite, while the fibres in the sample GPNS-1.0/FF and GPNS-3.0/FF did not show signs of significant damage after the ageing period (Fig. 5.28c, d). When compared this finding to our investigations in the case of incorporating nanoclay particles, the composites of flax fibres and geopolymer has experienced a reduction in their flexural strength by about 14% (Assaedi et al. 2017). In a comparable investigation, Aly et al. (2011) reported that durability was improved when nanoclay were added to cement mortar with waste glass. Furthermore, degradation of flax fibres was mitigated in the composites because alkalinity was reduced in the matrix. Another study found that when calcined nanoclay was added to cement composites that are reinforced using hemp fabric, the durability improved and the hemp fibres experienced less degradation (Hakamy et al. 2016). In yet another study, when meta-kaolin was added into mortar (that was reinforced using sisal fibre) at 28 days and following 25 wet and dry cycles, there was an improvement in durability (Filho et al. 2013). The meta-kaolin composites experienced a reduction in flexural strength by 23% in comparison to control composites at 28 days. Further, it was determined that sisal fibres in cement matrices did not experience significant degradation when 50% meta-kaolin was loaded. In the present study, the flax fibres degradation in the matrices was typically reduced, and the interfacial bond between the natural fibres and geopolymer nanocomposites was enhanced.

These desirable attributes are attractive for geopolymer composites and nanocomposites to be used in a variety of civil engineering structures. Possible applications for cotton fibre-reinforced geopolymer composites include slabs or shingles for siding, certain types of roofing, and some interior uses in the building structure. They may also be used for other applications such as pipes and cooling towers.

Acknowledgements The authors are grateful to Mr Andreas Viereckle of Mechanical Engineering at Curtin University for assistance with Charpy impact test. The authors would also like to thank Elaine Miller for her assistance with SEM.

References

Abanilla MA, Karbhari VM, Li Y (2006) Interlaminar and intralaminar durability characterization of wet layup carbon/epoxy used in external strengthening. Compos B 37:650–661

Alamri H, Low IM, Alothman Z (2012) Mechanical, thermal and microstructural characteristics of cellulose fibre reinforced epoxy/organoclay nanocomposites. Compos Part B: Eng 43:2762–2771

Al-Mosawi AI (2009) Study of some mechanical properties for polymeric composite material reinforced by fibers. Al-Qadisiya J Eng Sci 2:14–24

Aly M, Hashmi MSJ, Olabi AG, Messeiry M, Hussain AI, Abadir EF (2011) Effect of nano-clay and waste glass powder on the properties of flax fibre reinforced mortar. J Eng Appl Sci 6:19–28

Alzeer M, Mackenzie KD (2012) Synthesis and mechanical properties of new fibre-reinforced composites of inorganic polymers with natural wool fibres. J. Mater Sci 47:6958–6965

Alzeer M, Mackenzie K (2013) Synthesis and mechanical properties of novel composites of inorganic polymers (geopolymers) with unidirectional natural flax fibres (phormium tenax). Appl Clay Sci 75–76:148–152

Assaedi H, Shaikh FUA, Low IM (2016) Characterizations of flax fabric reinforced nanoclay-geopolymer composites. Compos B 95:412–422

Assaedi H, Shaikh FUA, Low IM (2017) Effect of nanoclay on durability and mechanical properties of flax fabric reinforced geopolymer composites. J Asian Ceram Soc 5:62–70

Assaedi H, Shaikh FUA, Low IM (2016a) Effect of nano-clay on mechanical and thermal properties of geopolymer. J Asian Ceram Soc 4:19–28

Assaedi H, Shaikh FUA, Low IM (2016b) Influence of mixing methods of nano silica on the microstructural and mechanical properties of flax fabric reinforced geopolymer composites. Constr Build Mater 123:541–552

Assaedi H, Alomayri T, Shaikh F, Low IM (2019) Influence of nano silica particles on durability of flax fabric reinforced geopolymer composites. Materials 12. https://doi.org/10.3390/ma12091459

Athijayamani M, Thiruchitrambalam U, Natarajan B (2009) Mater Sci Eng A 517:344–353

Azwa ZN, Yousif BF, Manalo AC, Karunase W (2013) A review on the degradability of polymeric composites base on natural fibres. Mater Des 47:424–442

Bakharev T (2006) The thermal behaviour of geopolymer prepared using class F fly ash and elevated temperature curing. Cem Concr Res 36:1134–1147

Bentur A, Mindess S (2007) Fibre reinforced cementitious composites, 2nd edn. Taylor & Francis, New York, pp 456–458

Chindaprasirt P, Jaturapitakkul C, Chalee W, Rattanasak U (2009) Comparative study on the characteristics of fly ash and bottom ash geopolymers. Waste Manage 29:539–543

Davidovits J (1991) Geopolymers—inorganic polymeric new materials. J Therm Anal 37(2002):1633–1656

de A Melo Filho J, de A Silva F, Toledo Filho RD (2013) Degradation kinetics and aging mechanisms on sisal fiber cement composite systems. Cem Concr Compos 40:30–39

de Andrade Silva F, Mobasher B, Toledo Filho RD (2009) Cracking mechanisms in durable sisal fiber reinforced cement composites. Cem Concr Compos 31:721–730

Deer WA, Howie RA (1996) An introduction to the rock-forming minerals, 2nd edn. Essex, England

Dhakal HN, Zhang ZY, Richardson MOW (2007) Effect of water absorption on the mechanical properties of hemp fibre reinforced unsaturated polyester composites. Compos Sci Technol 67:1674–1683

Dhakal HN, Zhang ZY, Bennett N, Arraiza AL, Vallejo FJ (2013) Effects of water immersion ageing on the mechanical properties of flax and jute fibre biocomposites evaluated by nanoindentation and flexural testing. J Compos Mater 48:1399–1406

Duxson P, Fernández-Jiménez A, Provis JL, Lukey GC, Palomo A, Van Deventer JSJ (2007) Geopolymer technology: the current state of the art. J Mater Sci 42:2917–2933

Espert A, Vilaplana F, Karlsson S (2004) Comparison of water absorption in natural cellulosic fibres from wood and one-year crops in polypropylene composites and its influence on their mechanical properties. Compos Part A 35:1267–1276

Filho RDT, Ghavami K, England GL, Scrivener K (2003) Development of vegetable fibre – mortar composites of improved durability. Cem Concr Compos 25:185–196

Filho JD, Silva DA, Toledo RD (2013) Degradation kinetics and aging mechanisms on sisal fibre cement composite systems. Cem Concr Compos 40:30–39

Ghosh R, Reena G, Krishna AR, Raju BL (2011) Effect of fibre volume fraction on the tensile strength of Banana fibre reinforced vinyl ester resin composites. Int J Adv Eng Sci Technol 4:90–91

Girisha C, Sanjeevamurthy GR, Srinivas K (2012) Sisal/coconut coir natural fibers-epoxy composites: Water absorption and mechanical properties. Int J Eng Innov Technol 2:166–170

Graupner N (2008) Application of lignin as natural adhesion promoter in cotton fibre-reinforced poly (lactic acid) (PLA) composites. J Mater Sci 43:5222–5229

Hakamy A, Shaikh FUA, Low IM (2015) Characteristics of nanoclay and calcined nanoclay-cement nanocomposites. Compos B 78:174–184

Hakamy A, Shaikh FUA, Low IM (2016) Effect of calcined nanoclay on the durability of NaOH treated hemp fabric-reinforced cement nanocomposites. Mater Des 92:659–666

Jo BW, Kim CH, Tae G, Park JB (2007) Characteristics of cement mortar with nano-SiO_2 particles. Constr Build Mater 21:1351–1355

Karmaker AC (1997) J Mater Sci Lett 16:460–462

Kim HJ, Seo DW (2006) Effect of water absorption fatigue on mechanical properties of sisal textile-reinforced composites. Int J Fatigue 28:1307–1314

Lau D, Büyüköztürk O (2010) Fracture characterization of concrete/epoxy interface affected by moisture. Mech Mater 42:1031–1042

Lee BH, Kim HJ, Yu WR (2009) Fabrication of long and discontinuous natural fiber reinforced polypropylene biocomposites and their mechanical properties. Fib Polym 10:83–90

Li W, Xu J (2009) Mechanical properties of basalt fibre reinforced geopolymeric concrete under impact loading. Mater Sci Eng 505:178–186

Li Q, Xu H, Li F, Li P, Shen L, Zhai J (2012) Synthesis of geopolymer composites from blends of CFBC fly and bottom ashes. Fuel 97:366–372

Liew YM, Heah CY, Mohd Mustafa AB, Kamarudin H (2016) Structure and properties of clay-based geopolymer cements: a review. Progr Mater Sci 83:595–629

Lin T, Jia D, He P, Wang M, Liang D (2008) Effects of fiber length on mechanical properties and fracture behavior of short carbon fiber reinforced geopolymer matrix composites. Mater Sci Eng A 497:181–185

Liu W, Ho SV, Pugh M (2005) Fracture toughness and water uptake of high-performance epoxy/nanoclay nanocomposites. Compos Sci Technol 65:2364–2673

Low IM, Mcgrath M, Lawrence D, Schmidt P, Lane J, Latella BA, Sim KS (2007) Mechanical and fracture properties of cellulose-fibre-reinforced epoxy laminates. Compos Part A 38:963–974

Ma X, Yu J, Kennedy JF (2005) Studies on the properties of natural fibers oreinforced thermoplastic starch composites. Carbohydr Polym 62:19–24

Maggana C, Pissis P (1999) Water sorption and diffusion studies in an epoxy resin system. J Polym Sci Pol Phys 37:1165–1182

Mishra S, Mohanty AK, Drzal LT, Misra M, Parija S, Nayak SK, Tripathy SS (2003) Studies on mechanical performance of biofiber/glass reinforced polyester hybrid composites. Compos Sci Technol 63:1377–1385

Mohamed AM (2016) Influence of nano materials on flexural behavior and compressive strength of concrete. HBRC J 12:212–225

Mohan TP, Kanny K (2011) Water barrier properties of nanoclay filled sisal fibre reinforced epoxy composites. Compos Part A 42:385–393

Mohanty AK, Misra M, Drazal LT (2005) Natural fibers, biopolymers, and biocomposites. CRC Press, Boca Raton

Mohr BJ, Biernacki JJ, Kurtis KE (2006) Microstructural and chemical effects of wet/dry cycling on pulp fiber-cement composites. Cem Concr Res 36:1240–1251

Mohr BJ, Biernacki JJ, Kurtis KE (2007) Supplementary cementitious materials for mitigating degradation of kraft pulp fibre-cement composites. Cem Concr Res 37:1531–1543

Nazari A, Sanjayan JG (2015) Hybrid effects of alumina and silica nanoparticles on water absorption of geopolymers: application of Taguchi approach. Measurement 60:240–246

Pacheco-Torgal F, Jalali S (2011) Cementitious building materials reinforced with vegetable fibres: a review. Constr Build Mater 25:575–581

Pernica D, Reis PNB, Ferreira JAM, Louda P (2010) Effect of test conditions on the bending strength of a geopolymer-reinforced composite. J Mater Sci 45:744–749

Phair JW, Van Deventer JSJ (2001) Eect of silicate activator pH on the leaching and material characteristics of waste-based inorganic polymers. Min Eng 14:289–304

Phair JW, Van Deventer JSJ (2002) Effect of the silicate activator pH on the microstructural characteristics of waste-based geopolymers. Int J Miner Process 66:121–143

Phoo-ngernkham T, Chindaprasirt P, Sata V, Hanjitsuwan S, Hatanaka S (2014) The effect of adding nano-SiO_2 and nano-Al_2O_3 on properties of high calcium fly ash geopolymer cured at ambient temperature. Mater Des 55:58–65

Pickering KL (2008) Properties and performance of natural-fibre composites. Woodhead Publishing, Cambridge

Rattanasak U, Chindaprasirt P (2009) Influence of NaOH solution on the synthesis of fly ash geopolymer. Miner Eng 22:1073–1078

Ray D, Rout J (2005) In: Thermoset biocomposites. Mohanty AK, Misra M, Drzal LT (eds) Taylor & Francis Group, New York, pp 310–313

Reddy CR, Sardashti AP, Simon LC (2010) Compos Sci Technol 70:1674–1680

Reis JML (2006) Fracture and flexural characterization of natural fiber-reinforced polymer concrete. Constr Build Mater 20:673–678

Reis JML (2006) Fracture and flexural characterization of natural fiber-reinforced polymer concrete. Constr Build Mater 20:73–78

Ren D, Yan C, Duan P, Zhang Z, Li L, Yan Z (2017) Durability performances of wollastonite, tremolite and basalt fiber-reinforced metakaolin geopolymer composites under sulfate and chloride attack. Constr Build Mater 134:56–66. https://doi.org/10.1016/j.conbuildmat.2016.12.103

Rong Z, Sun W, Xiao H, Jiang G (2015) Effects of nano-SiO_2 particles on the mechanical and microstructural properties of ultra high-performance cementitious composites. Cem Concr Compos 56:25–31

Sabinesh S, Thomas Renald CJ, Sathish S (2014) Compos Sci Technol 65:805–813

Schneider H, Schreuer J, Hildmann B (2008) Structure and properties of mullite—a review. J Eur Ceram Soc 28:329–344

Shaikh FUA (2013) Deflection hardening behaviour of short fibre reinforced fly ash based geopolymer composites. Mater Des 50:674–682

Silva FA, Toledo Filho RD, Melo Filho JA, Fairbairn EMR (2010) Physical and mechanical properties of durable sisal fiberecement composites. Constr Build Mater 24(5):777–785

Supit SWM, Shaikh FUA (2014) Durability properties of high-volume fly ash concrete containing nano-silica. Mater Struct 48:2431–2445

Toledo Filho RD, Scrivener K, England GL, Ghavami K (2000) Durability of alkali-sensitive sisal and coconut fibres in cement mortar composites. Cem Concr Compos 22:127–143

Ul Haq E, Kunjalukkal Padmanabhan S, Licciulli A (2014) Synthesis and characteristics of fly ash and bottom ash based geopolymers—a comparative study. Ceram Int 40:2965–2971

Vijai K, Kumuthaa R, Vishnuram BG (2012) Properties of glass fibre reinforced geopolymer concrete composites. Asian J Civ Eng 13:511–520

Wambua P, Ivens J, Verpoest I (2003) Natural fibres: can they replace glass in fibre reinforced plastics? Compos Sci Tech 63:1259–1264

Yadollahi MM, Benli A, Demirboğa R (2015) The effects of silica modulus and aging on compressive strength of pumice-based geopolymer composites. Constr Build Mater 94:767–774

Yan L, Chouw N (2015) Effect of water, seawater and alkaline solution ageing on mechanical properties of flax fabric/epoxy composites used for civil engineering applications. Constr Build Mater 99:118–127

Zaharaki D, Komnitsas K, Perdikatsis V (2010) Use of analytical techniques for identification of inorganic polymer gel composition. J Mater Sci 45:2715–2724

Zhang YJ, Wang YC, Xu DL, Li S (2010) Mechanical performance and hydration mechanism of geopolymer composite reinforced by resin. Mater Sci Eng A 527:6574–6580

Chapter 6
Thermal Stability and Flammability

Abstract This chapter describes the thermal characteristics and flammability of geopolymer composites reinforced with natural fibres and/or nanofillers. Results show that the appropriate addition of natural fibres and/or nanofillers can a significant influence on the thermal stability and flammability of these materials.

6.1 Cotton Fabric Reinforced Geopolymer Composites

6.1.1 Effect of Elevated Temperature

The thermal stability of the composites was studied using the thermogravimetric analysis (TGA); thermograms of pure geopolymers, geopolymer composites and cotton fibres are shown in Fig. 6.1. The geopolymer matrix undergoes weight loss between 50 and 250 °C due to the evaporation of free water. The composites show a similar trend, with a slight additional weight loss above 250 °C which may be attributed to the thermal decomposition of the cotton fibres.

As shown in Fig. 6.1, pure cotton fibres degrade in three main stages. The first transition occurs from 50 °C to about 250 °C, with the release of absorbed moisture from the fibre by evaporation. In this stage, the weight loss of cotton fibres is roughly like that of other natural fibres such as kenaf, jute and wood (Ray et al. 2002; Huda et al. 2006; Julkapli and Akil 2010). The second transition occurs between 250 and 370 °C when a large weight loss occurs, attributable to the degradation of cellulose. The third stage occurs above 370 °C when the fibres start to decompose but shows a lower rate of weight loss. In this stage, all volatile materials are driven off, resulting in the formation of residual char in the sample. Similar degradation stages of cotton fibres were observed by Babu et al. (2009) when they studied the thermal decomposition characteristics of cotton fibres using the thermogravimetric analysis.

Figure 6.2 shows that the shrinkage of geopolymer composites increases with an increase of cotton fibres in the matrix. This tendency is observed for all samples at various temperatures. In all the three composites containing cotton fibres (4.5, 6.2 and 8.2 wt%), the drying shrinkage increases as temperature increases.

I.-M. Low et al., *Cotton and Flax Fibre-Reinforced Geopolymer Composites*, Composites Science and Technology,
https://doi.org/10.1007/978-981-16-2281-6_6

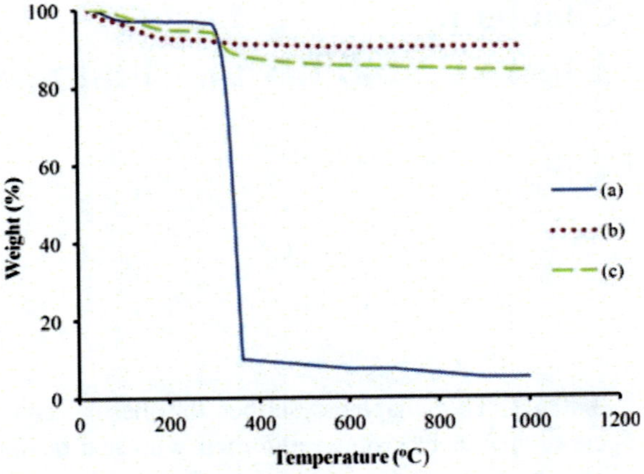

Fig. 6.1 TGA curves of **a** cotton fibres, **b** pure geopolymers and **c** geopolymer composites

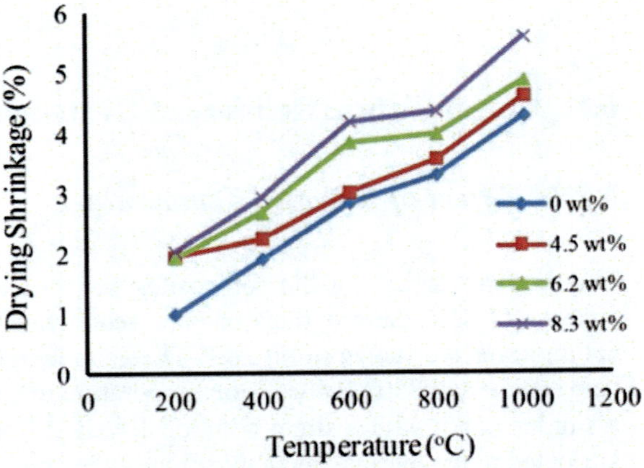

Fig. 6.2 Drying shrinkage of geopolymer composites at various temperatures

This is in contrast with reports of reduced shrinkage in geopolymer composites containing a small quantity of synthetic fibres such as carbon at elevated temperatures (Lin et al. 2009). Shrinkage of geopolymer composites is influenced by many factors, including fibre type and fibre content when drying begins. The results here indicate that the addition of cotton fibres increases the matrix porosity (Fig. 6.3) and contributes to a higher drying shrinkage. This is due to more moisture paths being created into the matrix, which is also reported by Toledo Filho et al. (2005) in their study of the effect of vegetable fibres on drying shrinkage of cement mortar composites.

After heating at various temperatures, the samples were assessed visually to determine the color change. Photographs of the geopolymer composite samples before and after heating at various temperatures are shown in Fig. 6.4.

All samples exhibit a change in color after exposure to elevated temperatures as follows: grey at 200 °C, yellowish grey between 400 and 600 °C, pink up to 800 °C, and reddish brown at 1000 °C. High temperature exposure causes the oxidation and

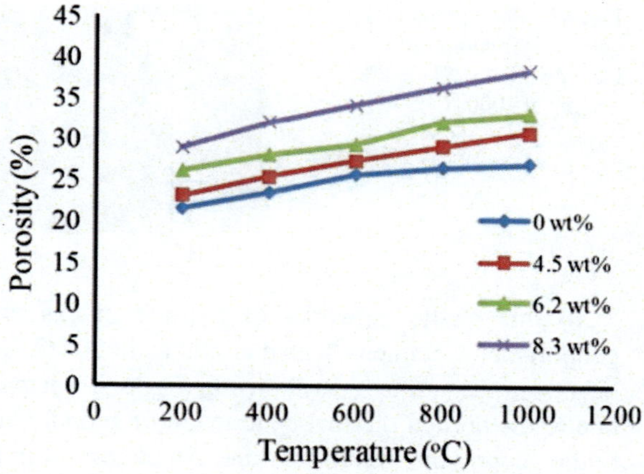

Fig. 6.3 Porosity of geopolymer composites at various temperatures

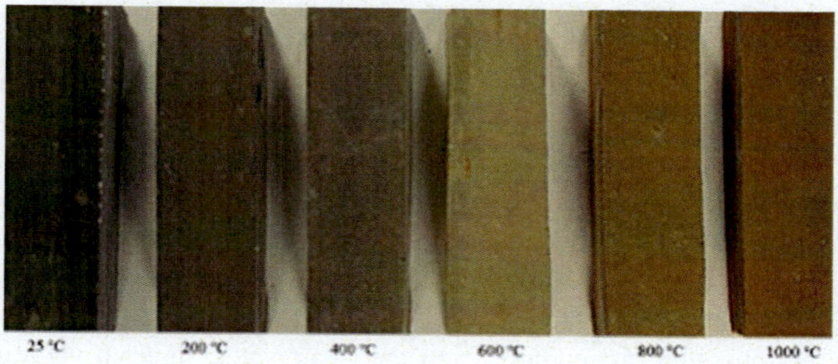

Fig. 6.4 Geopolymer composites loaded with 8.3 wt% CF before and after heating at various temperatures

liberation of iron species present in the fly-ash particles, where similar observations were also reported by Rickard et al. (2010) when testing the color change of Collie fly-ash geopolymers at elevated temperatures. Such color changes are a useful tool for estimating temperatures easily which have been reached after exposure to fire. They can also be used as an indication of significant loss in mechanical properties and are useful since the appearance coincides with a significant reduction in strength because of heating (Short et al. 2001).

It is well known that cracks occur in concrete during fire because of the internal pressure of evaporable moisture, characterised by hairline cracks over the surface at low to medium temperatures and extending deep into the specimen at higher temperatures. The result is a significant reduction in structural integrity and load-carrying capacity (Şahmaran et al. 2011). One of the primary mechanisms responsible for concrete cracking is low porosity, which slows the migration of water vapour and offers few escape channels, leading to higher pore pressures (Khaliq and Kodur 2011).

Fig. 6.5 Photograph of the geopolymer composites loaded with 0 wt% CF after heating at 1000 °C

In this study, networks of hairline cracks are observed on un-reinforced geopolymer specimens heated at 400 and 600 °C. Beyond 600 °C, severe cracking occurs on the surfaces, as shown in Fig. 6.5. It is probable that cracking is initially due to the normal thermal contraction of geopolymer paste, causing local changes to the geopolymer microstructure. An increment in temperature leads to additional contraction in the matrix as water is driven off. This phenomenon has been observed to occur in the geopolymer gel during dehydroxylation between 300 and 600 °C (Duxson et al. 2007). However, when cotton fibres are incorporated, no crack is found in the surface of geopolymer composites as shown in Fig. 6.4. This indicates that cotton fibres are very effective in prevention of cracking caused by high temperatures due to the formation of small cavities inside the matrix created by the fibre degradation. The structure of the composites becomes more porous, and the expanded water vapour escapes without substantial damage to the microstructure.

Such degradation of the cotton fibres may be beneficial to the behavior of geopolymer composites under thermal exposure. At high temperatures, when all water is not expelled fast enough from the composites, internal vaporisation may create high pressures inside the matrix. The porosity and small channels created by the degradation of the cotton fibres may lower internal vapour pressures and thus reduce the likelihood of cracking.

Microstructure Analysis

To observe changes in microstructure at moderate length scales, microscopy at low magnification and with a large field of view was conducted. Optical images in Fig. 6.6a–c show the micrographs of specimens exposed to elevated temperatures from 200 to 600 °C. A significant amount of oxidation appears as dark "burns out" areas along the outer perimeter of the samples; but the centres remain pristine because oxygen is quickly depleted at the edge of the sample and no oxygen remains to react with the cotton fibres inside. The good condition of the interior fibres indicates that the geopolymer paste aids in stopping or significantly decreasing the supply of oxygen to the interior of the composites.

The matrix is effective in cutting off the oxygen supply to the fibres at temperature up to 600°C (Fig. 6.6a–c); however, as the specimen is heated further, many voids and caverns form around the edge and along the matrix (Fig. 6.6d–e), probably owing to the severe degradation of cotton fibres caused by oxidation along the edge and deep into the interior of the sample. As the temperature approaches 800 and 1000 °C, oxygen is better able to diffuse into the interior and react with the fibres, leading to

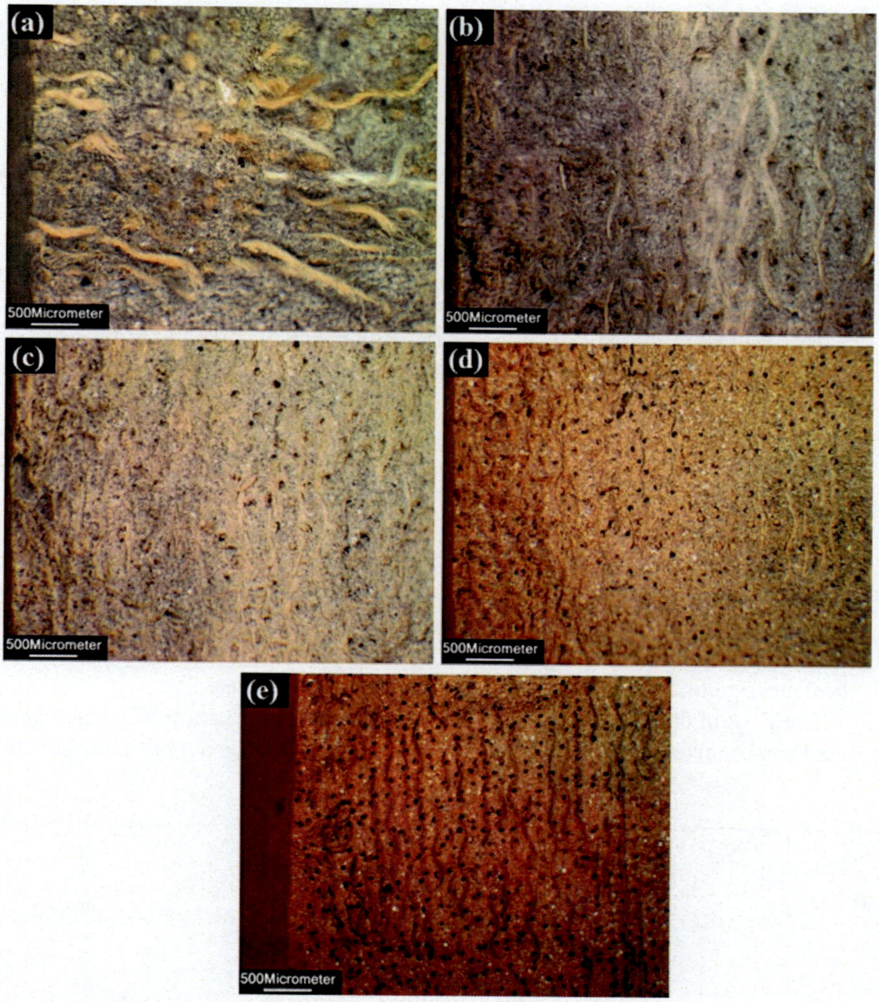

Fig. 6.6 Optical micrographs of geopolymer composites loaded with 8.3 wt% CF at various temperatures: **a** 200 °C, **b** 400 °C, **c** 600 °C, **d** 800 °C and **e** 1000 °C

complete fibre burnout. However, no crack is observed in any of the specimens at these temperatures, possibly because the degrading cotton fibres form the dehydration pathways along which water could escape, preventing the build-up of pore pressures.

Therefore, the addition of cotton fibres to a geopolymer matrix has a positive influence, enhancing the temperature resistance and the mechanical and fracture properties of the composites. This makes them suitable for use in technological applications which are likely to be exposed to extreme high temperatures during wildfire such as underground structures and reinforcement of mine works (e.g. adits, tunnels, etc.), as railroad and road tunnel constructions are.

6.1.2 Effect of Nanoclay

Thermal Stability

The thermal stability of samples was determined using thermogravimetric analysis (TGA). In this test, the thermal stability was studied in terms of the weight loss percentage as a function of temperature in Argon atmosphere. The results of thermogravimetric analysis (TGA) and the differential thermogravimetry (DTG) of all samples are shown in Figs. 6.7 and 6.8, respectively. The residual mass at different temperatures for nano-clay and all samples are summarized in Table 6.1.

The residual mass results exhibit the remaining weight percentage of the material after The TGA test, which could be an indicator of the organic/inorganic component of the material (Ul Haq et al. 2014). For instance, in the case of nano-clay, TGA showed that about 30 wt% of the component burnt at 800 °C, which is the organic component in the Cloisite 30 B. This is equivalent to the loss on ignition wt% of the supplier technical data sheet of the Cloisite 30 B.

Figure 6.7 shows that the major decomposition region of the nano-clay is between 300 and 450 °C, where the weight loss is about 21%. However, there are no noteworthy changes occurred to the weight loss of the nanocomposite curves at the same region, rather they showed almost the same weight loss of the pure geopolymer. This shows that the decomposition of nano-clay did not influence the thermal stability of the nanocomposite due to the small wt% nano-clay additions.

The TGA of the pure geopolymer and the nanocomposites showed a major weight loss from room temperature to 150 °C, due to the evaporation of physically adsorbed

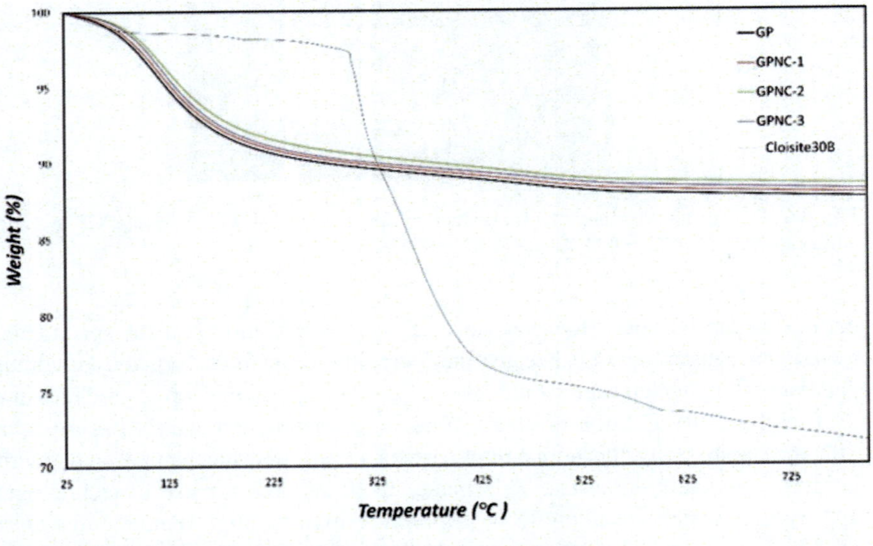

Fig. 6.7 TGA curves of nano-clay (Cloisite 30B) and all samples

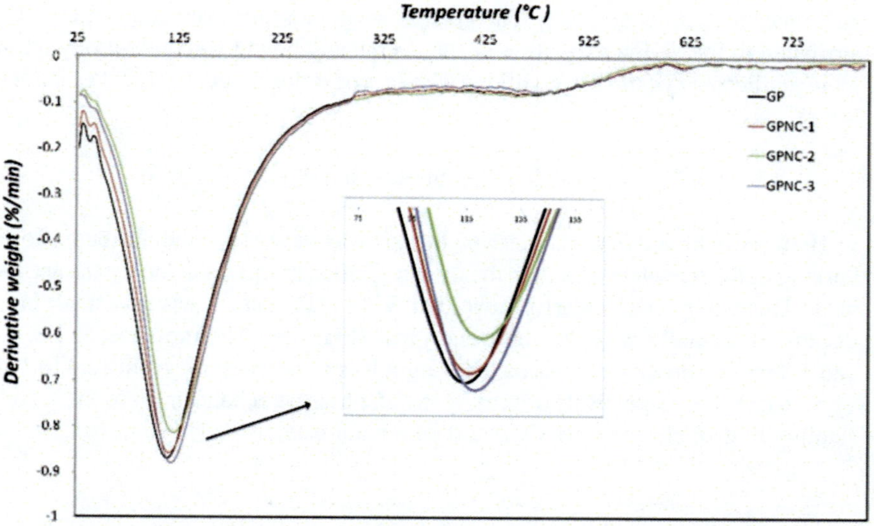

Fig. 6.8 DTG curves of pure geopolymer and geopolymer nanocomp

Table 6.1 Thermal properties of control paste, geopolymer nanocomposites and nano-clay (Cloisite 30B)

Sample	Residual mass at different temperatures (%)							
	100 °C	200 °C	300 °C	400 °C	500 °C	600 °C	700 °C	800 °C
GP	97.1	91.1	89.7	89.0	88.6	87.8	87.7	87.5
GPNC-1	97.5	91.4	90.0	89.2	88.5	88.1	88.0	87.8
GPNC-2	98.1	92.1	90.6	89.8	89.0	88.7	88.6	88.4
GPNC-3	97.8	91.6	90.1	89.4	88.8	88.4	88.2	88.1
Nano-clay	98.6	98.2	97.3	78.7	75.4	73.5	72.6	71.5

water. The neat geopolymer curve is steeper in this region compared to the nanocomposites curves, which is clearly shown in DTG graph (Fig. 6.8) where the peak of pure geopolymer moved to a lower temperature compared to the nanocomposite. This is due to the formation of high porosity in the neat geopolymer, which reduces the ability of the sample to retain water. On the other hand, GPNC-2 sample exhibited the highest thermal stability among all geopolymer samples, the peak of the nanocomposite containing 2.0 wt% nano-clay shifted slightly to a higher temperature than other samples. This may be attributed to the effect of nano-clay filling the voids, producing denser geopolymer, and/or it may be attributed to the nano-composite specimens had higher amounts of geopolymer gel and amorphous content. Between 150 and 300 °C, the rate of weight loss for all samples started to slow as the physical free water had already evaporated, and the interstitial water (Zaharaki et al. 2010) started to decompose, this is the water molecules that were possibly associated with sodium cations (Li et al. 2012). The gradual weight loss between 300 and 600 °C is attributed to the de-hydroxylation of the chemically bound silicon-hydroxyl group giving (silicon-oxygen-silicon) bridge with loss of water (Li et al. 2012; Ferone et al. 2013).

$$2Si{-}OH \longrightarrow Si{-}O{-}Si + H_2O\uparrow$$

Between 600 and 700 °C the weight loss was very slow and attributed to the burning of the remnants of coal in the fly ash. Carbon remnants in the fly ash are 1.45 wt%. This is clear particularly above 600 °C in DTG curves where a small hump displaying a small change of the weight loss (Fig. 6.8). The presence of 1.0, 2.0, and 3.0 wt% nano-clay decreased the weight loss of geopolymer from 12.4 to 12.1, 11.5 and 11.8%, respectively, revealing that the highest enhancement to the thermal stability of geopolymer matrix was 2.0 wt% nano-clay.

6.1.3 *Effect of Ordinary Portland Cement*

Thermal Stability

The thermal stability of the samples was determined using thermogravimetric analysis (TGA). In this test, thermal stability was studied in terms of weight loss as a function of temperature under atmospheric conditions. The thermograms (TGA) of the geopolymer matrix and CF-reinforced geopolymer composite are shown in Figs. 6.9 and 6.10.

In general, the thermogram of pure geopolymer shows weight loss from 50 to 300 °C, because of the dehydration of physically adsorbed water. Above 300 °C, the loss is attributed to the dehydroxylation of the chemically bound water. The weight

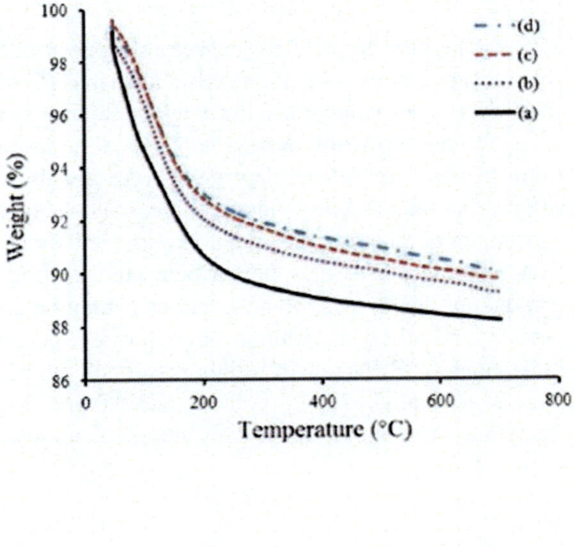

Fig. 6.9 TGA curves of geopolymer matrix containing **a** 0 wt% OPC; **b** 5 wt% OPC; **c** 8 wt% OPC and **d** 10 wt% OPC

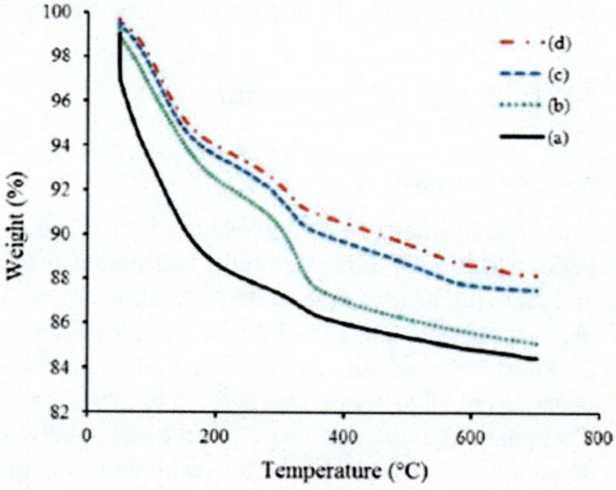

Fig. 6.10 TGA curves of CF-reinforced geopolymer composites containing **a** 0 wt% OPC; **b** 5 wt% OPC; **c** 8 wt% OPC and **d** 10 wt% OPC

loss of neat geopolymer decreases as a result of the presence of OPC (see Fig. 6.9). The presence of 5, 8 and 10 wt% OPC decreases the weight loss for geopolymer from 11.16 to 10.2, 9. 11 and 9.06 °C, respectively, revealing that the addition of OPC decreases moisture access to the sample. This indicates that thermal stability enhances as OPC content increases for geopolymer samples. In a similar study, Chen et al. (2012) reported that the addition of 10 wt% nano-TiO_2 into cement paste improved the thermal stability of nanocomposites.

The addition of OPC to CF-reinforced geopolymer composite and its influence on thermal stability can be seen in Fig. 6.10. The CF-reinforced geopolymer composite containing OPC showed lower weight loss and better thermal stability than those without OPC. Compared to the weight loss of 10% for CF/geopolymer without OPC, the weight loss of CF-reinforced geopolymer composite with 10 wt% OPC was only 4.64% at 200 °C. This enhancement of thermal properties is the result of better penetration of the OPC between the cotton fabric openings, resulting in a strong fibre–matrix interface and an increase in thermal stability. In addition, the presence of OPC in geopolymer composites may provide a thermal barrier which reduces the ingress of air and the resultant oxidative degradation; therefore, the resistance to thermal degradation of cotton fibre reinforced composites can be improved by the addition of OPC.

6.2 Flax Fabric Reinforced Geopolymer Composites

6.2.1 Effect of Composition

Thermal Stability

The thermal stability of samples was determined using thermogravimetric analysis (TGA). In this test, thermal stability was studied in terms of the weight loss percentage as a function of temperature in argon atmosphere. The thermograms (TGA) of FF, neat geopolymer and FF-reinforced geopolymer composite are shown in Fig. 6.11.

The thermogram of flax fibres shows degradation in three steps. The first transition occurs from 25 to approximately 240 °C, with the release of free water evaporation. Then, the largest weight loss (%) occurred at between 240 and 365 °C is due to the decomposition of cellulose. This result agrees with Alzeer and MacKenzie (2013), where the highest weight loss (%) of short flax fibres under flowing air is in the range 240–340 °C. The final stage occurs above 365 °C, when the fibres start to decompose but displays a lower rate of weight loss, and all volatile substances are dispelled.

The pure geopolymer shows weight loss occurring from 25 to 300 °C, caused by the evaporation of physically adsorbed water. Above 300 °C, weight loss is attributed to the dehydroxylation of the chemically bound water. The FF–reinforced geopolymer composite shows a weight loss of 10.5% up to about 260 °C, which is due to the evaporation of physically absorbed water. Above 260 °C the composite shows further weight loss because of the degradation of the fibre content in the sample. The porosity of geopolymer matrix allows the oxygen to enter and cause degradation of the flax fibres at high temperatures. The composite shows a total weight loss of ~15% at 300°C which indicates further degradation of fibres inside the composite. At

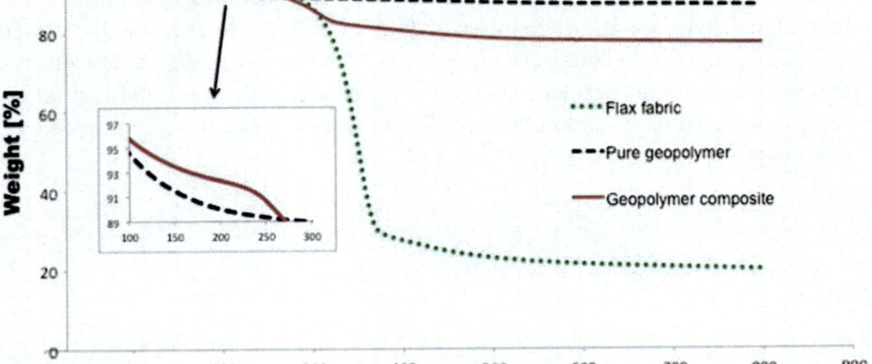

Fig. 6.11 TGA curves of the flax-reinforced geopolymer composite, the matrix and the flax fibres

this temperature a substantial amount of fibre degradation had occurred. Therefore, it could be concluded that this composite system is only suitable for service up to 250 °C. It is worth mentioning here that the TGA micro-sample is not necessarily representative of the whole composite sample because the distribution of flax fibres is not uniform within the geopolymer matrix. Therefore, the fibre content of the TGA micro-sample will be highly dependent on the position it is taken from the composite sample. However, TGA test can provide a good estimation of the thermal stability of a composite when compared to the thermal stability of its components.

6.2.2 *Effect of Nanoclay*

Thermal Behaviour

The thermal stability of neat geopolymer and geopolymer nanocomposites was analyzed using thermogravimetric analysis (TGA) and derivative thermograms (DTG). In this investigation, the thermal stability was studied in terms of the weight-loss percentage as a function of temperature in Argon atmosphere. The results are shown in Fig. 6.12.

The thermograms of the pure geopolymer and the nanocomposites samples display a weight loss from 25 to 225 °C due to the evaporation of absorbed water (Zivica et al. 2014). The neat geopolymer curve shows sharp decrease in this region compared to the nanocomposite curves, which is clearly shown in DTG graph, where the peak

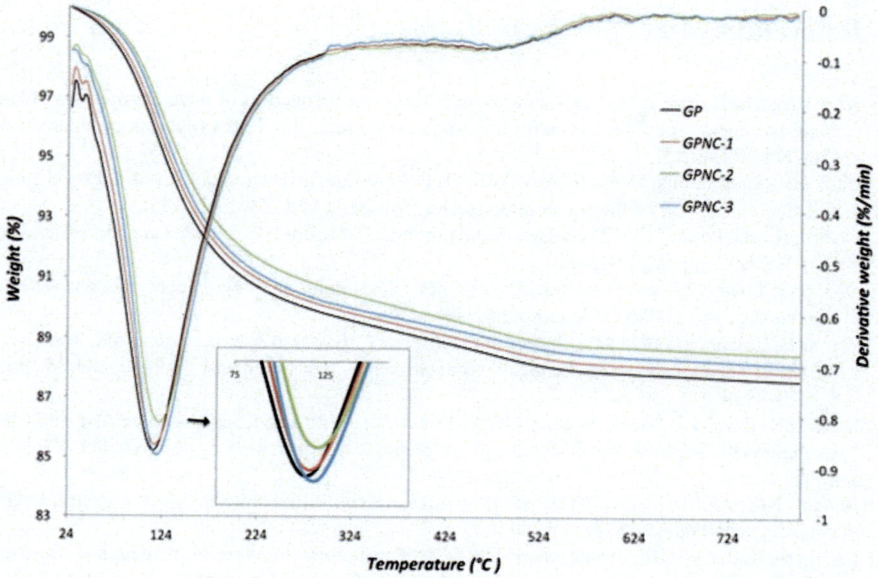

Fig. 6.12 TGA/DTG curves of pure geopolymer and geopolymer nanocomposites

of nanocomposites shifted to higher temperatures compared to the neat geopolymer. GPNC-2 shows the lowest reduction of rate of weight loss indicating that geopolymer containing 2.0 wt% nanoclay has the lowest water content compared to the tested samples. This may be attributed to the effect of nanoclay filling the voids, producing denser matrices. Between 225 and 525 °C, the rate of weight loss for all samples was slow as the physical free water was evaporated. This gradual weight loss is recognized as the de-hydroxylation of the chemically bound silicon-hydroxyl group giving silicon-oxygen group and evaporated water (Li et al. 2012).

$$2Si{-}OH \longrightarrow Si{-}O{-}Si + H_2O \uparrow$$

Between 500 and 700 °C the weight loss was slow and attributed to the burning of the remaining coal of fly ash (Ul Haq et al. 2014). This is clear specifically above 600 °C in DTG curves where a small hump displaying a small change of the weight loss. The presence of 1.0, 2.0, and 3.0 wt% nanoclay reduced the weight loss of geopolymer from 12.4 to 12.1%, 11.5 and 11.8%, respectively, revealing that the highest improvement to thermal stability of geopolymer matrix was 2.0 wt% loading of nanoclay. Note that composites reinforced with flax fibres have not been investigated. This is because that TGA technique is very sensitive to the fibre/matrix ratio, which cannot be fixed for all composites containing FF, considering the small weight of the TGA micro-samples and the flax fibres size. However, the main objective of this study is to determine the effect of nanoclay on the physical structure of the geopolymer/nanocomposites matrices.

References

Alzeer M, MacKenzie K (2013) Synthesis and mechanical properties of novel composites of inorganic polymers (geopolymers) with unidirectional natural flax fibres (phormium tenax). Appl Clay Sci 76:148–152

Babu KF, Senthilkumar R, Noel M et al (2009) Polypyrrole microstructure deposited by chemical and electrochemical methods on cotton fabrics. Synthetic Met 159:1353–1358

Chen J, Kou S, Poon C (2012) Hydration and properties of nano-TiO_2 blended cement composites. Cem Concr Compos 34:642–649

Duxson P, Lukey GC, van Deventer JSJ (2007) Physical evolution of Na-geopolymer derived from metakaolin up to 1000°C. J Mater Sci 42:3044–3054

Ferone C, Colangelo F, Roviello G, Asprone D, Menna C, Balsamo A, Prota A, Cioffi R, Manfredi G (2013) Application-oriented chemical optimization of a metakaolin based geopolymer. Materials 6:1920–1939

Huda MS, Drzal LT, Misra M et al (2006) Wood-fiber-reinforced poly(lactic acid) composites: Evaluation of the physicomechanical and morphological properties. J Appl Polym Sci 102:4856–4869

Julkapli NM, Akil HM (2010) Thermal properties of kenaf-filled chitosan biocomposites. Polym-Plast Technol 49:147–153

Khaliq W, Kodur V (2011) Thermal and mechanical properties of fiber reinforced high performance self-consolidating concrete at elevated temperatures. Cement Concrete Res 41:1112–1122

Li Q, Xu H, Li F, Li P, Shen L, Zhai J (2012) Synthesis of geopolymer composites from blends of CFBC fly and bottom ashes. Fuel 97:366–372
Lin T, Jia D, He P et al (2009) Thermal-mechanical properties of short carbon fiber reinforced geopolymer matrix composites subjected to thermal load. J Cent South Univ Tech 16:881–886
Ray D, Sarkar BK, Basak RK et al (2002) Study of the thermal behavior of alkali-treated jute fibres. J Appl Polym Sci 85:2594–2599
Rickard WDA, van Riessen A, Walls P (2010) Thermal character of geopolymers synthesized from class F fly ash containing high concentrations of iron and α-quartz. Int J Appl Ceram Tech 7:81–88
Şahmaran M, Özbay E, Yücel H et al (2011) Effect of fly ash and PVA fiber on microstructural damage and residual properties of engineered cementitious composites exposed to high temperatures. J Mater Civ Eng 23:1735–1745
Short NR, Purkiss JA, Guise SE (2001) Assessment of fire damaged concrete using colour image analysis. Constr Build Mater 15:9–15
Toledo Filho RD, Ghavami K, Sanjuán M et al (2005) Free, restrained and drying shrinkage of cement mortar composites reinforced with vegetable fibres. Cement Concrete Comp 27:537–546
Ul Haq E, Kunjalukkal Padmanabhan S, Licciulli A (2014) Synthesis and characteristics of fly ash and bottom ash based geopolymers—a comparative study. Ceram Int 40:2965–2971
Zaharaki D, Komnitsas K, Perdikatsis V (2010) Use of analytical techniques for identification of inorganic polymer gel composition. J Mater Sci 45:2715–2724
Zivica V, Palou MT, Bage TIL (2014) High strength metahalloysite based geopolymer. Compos Part B 57:155–165

Chapter 7
Summary and Concluding Remarks

Abstract This chapter provides a summary and concluding remarks on geopolymer composites reinforced with natural fibres and/or nanofillers. Potential applications, challenges, and future directions of these materials are highlighted and addressed.

7.1 Summary on Cotton Fibre Reinforced Geopolymer Composites

7.1.1 Short Cotton Fibre Reinforced Geopolymer Composites

Short cotton fibre-reinforced geopolymer composites with a range of fibre contents from 0 to 1.0 wt% were prepared by adding short cotton fibres in fly ash based geopolymer paste. The fabricated composites were investigated to identify and assess the effect of fibre reinforcement on the flexural modulus, flexural strength, impact strength, hardness, compressive strength and fracture toughness. It was found that the addition of cotton fibre reinforcements increased the flexural modulus, flexural strength and impact strength of geopolymer composites. For example, while pure geopolymer showed a flexural strength of 10.4 MPa, reinforcement with 0.3 and 0.5 wt% cotton fibre exhibited flexural strengths of 11.2 and 11.7 MPa, respectively.

The enhancements in flexural strength and flexural modulus of the geopolymer composites compared to geopolymer matrix are believed to be underpinned by good dispersion of cotton fibres throughout the matrix, which helps to increase the adhesion at the matrix/cotton fibre interface and permits the stress transfer from matrix to the fibres, resulting in an improvement of strength properties.

The addition of cotton fibre to 0.3 and 0.5 wt% was found to increase impact strength from 1.9 to 3.2 kJ/m^2 and 4.5 kJ/m^2, respectively. This enhancement in impact strength is believed to be due to the ability of the cotton fibres to transfer impact stress using energy dissipation mechanisms, primarily by fibre pullout, fibre fracture, and matrix deformation. Results were similar to compressive strength. Geopolymer

I.-M. Low et al., *Cotton and Flax Fibre-Reinforced Geopolymer Composites*, Composites Science and Technology,
https://doi.org/10.1007/978-981-16-2281-6_7

composites with 0.3 and 0.5 wt% cotton fibre were increased from 18 to 27 MPa and 45 MPa, respectively.

The addition of short cotton fibres was also improved the fracture toughness. For example, while the fracture toughness of pure geopolymer was recorded as 0.6 MPa.m$^{1/2}$, after the addition of 0.3 and 0.5 wt% cotton fibre, the fracture toughness was increased to 0.9 and 1.12 MPa.m$^{1/2}$, respectively. This improvement in fracture toughness is attributed to crack deflection, energy dissipation, and fracture resistance properties provided by the cotton fibres.

However, when fibre content increases to more than 0.5 wt% a different pattern emerged. The mechanical properties (flexural strength, flexural modulus, compressive strength, hardness, impact strength and fracture toughness) were not improved at this fibre content. It is believed that a high content of cotton fibres inhibits the inhomogeneity within the matrix so that agglomerations of fibres (confirmed with SEM) are formed, degrading interfacial adhesion between fibre and matrix. In addition, these agglomerations may act as stress concentrators, causing further reductions in flexural strength, flexural modulus, compressive strength, hardness, impact strength and fracture toughness. It was observed that increasing the content of cotton fibre caused a discernible increase in matrix viscosity. This was compensated for by increasing the water content of the mix and overcoming one problem in this way may have led to other adverse effects, such as an increase in porosity and microcracking. These usually lead to the reduction in bonding at the fibre–matrix interface, which results in lower stress transferred from the matrix to the fibres.

7.1.2 Cotton Fabric-Reinforced Geopolymer Composites

Geopolymer composites with different fibre loadings of cotton fabric (CF) were produced by hand-layup technique. The amount of binder varies as the quantity of fabric increases. Mechanical properties as function of fibre content were investigated. It was found that the increase in the cotton fibre content in geopolymer composites led to an enhancement of the mechanical properties of the composites. The flexural strength, flexural modulus, impact strength and fracture toughness are increased at an optimum fibre content of 2.1 wt% (three layers of cotton fabric). However, an increase in cotton fibre content beyond this level caused a reduction in mechanical properties, which is attributed to poor bonding between the fibre and matrix because of the higher porosity and the formation of voids trapped beneath the cotton fabric sheets during casting. With respect to thermal properties, it was found that the use of geopolymer binder improved the cotton fibre's resistance to thermal degradation. The geopolymer binder protects the cotton fibres, acting as barrier to reduce the ingress of air and resultant oxidative degradation.

The addition of Ordinary Portland Cement (OPC) enhanced the bonding, reduced the porosity and improved the mechanical properties of CF/geopolymer composites. By increasing the OPC content to 5, 8 and 10 wt% the flexural strength increased from 14.7 MPa for pure geopolymer to 16.2, 16.6 and 17.1 MPa, respectively. Similarly, the addition of OPC at 5, 8, and 10 wt% also increased the impact strength from 6.9 kJ/m^2 for pure geopolymer to 7.3, 7.5 and 7.8 kJ/m^2, respectively. The enhancement in flexural strength and impact strength of the CF/geopolymer composites with added OPC compared to those without is attributed to an enhancement in interfacial adhesion between fibre and matrix. This superior interfacial adhesion is believed to be caused by a filling effect, whereby the OPC particles fill voids or pores in the geopolymer paste making the microstructure of the matrix denser than that of the paste without OPC, and also by a pozzolanic reaction, in which the OPC with fly ash produces C-S-H gel in the geopolymeric system (confirmed by XRD and SEM), and reduces porosity, thereby enhancing mechanical properties.

When OPC content increased beyond 5 wt% the fracture toughness is decreased. It has already been considered that the OPC addition results in desirable strength properties as a result of an improvement in fibre–matrix adhesion; but this also makes the composite brittle, as indicated by the lower fracture toughness in all the samples. The addition of OPC causes fibre–matrix adhesion to be high but hinders the energy absorption mechanisms provided by fibre pullout and fibre de-bonding. For example, in composites with 10 wt% OPC, the SEM confirmed that cotton fibres were ruptured rather than pulled out.

With respect to thermal stability, the CF-reinforced geopolymer composites containing OPC showed lower weight loss and better thermal stability than those without OPC. This enhancement of thermal properties is the result of the formation of calcium silicate hydrate (CSH). In addition, the presence of OPC in geopolymer composites was found to provide a thermal barrier reducing the ingress of air and inhibiting oxidative degradation; thus, the addition of OPC to cotton fibre-reinforced composites was found to improve the fibre's resistance to thermal degradation.

The forced impregnation-hand layup process, which involved rolling and brushing the woven cotton fabric with geopolymer paste and then stacking the fabric/paste layers to a required thickness, was found to be the most effective method of making multiple layers while ensuring complete penetration of the matrix into the fibre bundles. This impregnation method was found to lead to the fabrication of composites with lower void content and good fibre–matrix adhesion. Several following conclusions were derived from the experiment.

- The flexural strength and modulus of geopolymer composites increase with the addition of woven cotton fibres, compared to pure geopolymer. It was observed that at peak load none of the specimens was completely broken. This could be because the crack bridging by long continuous fibres caused the flexural strength and modulus of woven cotton fibre-reinforced geopolymer composites to be higher than pure geopolymer. The flexural strength of geopolymer composites increased from 8.2 to 31.7 MPa as fibre content increased from 0 to 8.3 wt%; these enhancements are attributed to the ability of the fibres to resist shear failure and contribute in sustaining the applied load to the composites. The resistance of shear failure permits greater stress transfer between matrix and cotton fibres, resulting in improved flexural strength of the composites.

- The impact strength of the composites also increases as the cotton fibre content increases. The impact strength of the geopolymer is increased from 2.1 to 15.6 kJ/m^2 after the addition of 8.3 wt% cotton fibres. This improvement in impact strength may be attributed to the fibrillation of cotton layers creating branches in the fibre (confirmed by SEM), resulting in the formation of microfibrils which increase the specific surface area of the fibre. Increasing the surface area leads to enhanced fibre–matrix interaction and bonding between the microfibrils and the geopolymer matrix; thus, increased fibre–matrix bonding leads to superior stress transfer from the matrix to the microfibrils.
- Fracture toughness increases with higher fibre content. The composites with higher cotton fibre content exhibited higher fracture toughness. The greatest improvement in fracture toughness (from about 0.6 MPa $m^{1/2}$ for the unreinforced matrix to about 1.8 MPa $m^{1/2}$) was obtained with 8.3 wt% cotton fibre reinforcement. This enhancement is because of the unique properties of woven cotton fibre to resist fracture through increased energy dissipation from crack-deflection at the fibre–matrix interface, fibre-debonding, fibre-bridging, fibre pullout and fibre-fracture.
- The hardness of composites increases with higher fibre loading. This is because the uniform distribution of the load on cotton fibres decreases the penetration of the test ball to the surface of the composite and consequently increases the hardness of composite.
- The compressive strength of composites increases with the addition of cotton fibres. From 19.8 MPa for pure geopolymer, reinforcement with 8.3 wt% CF increased compressive strength to 90 MPa. This is attributed to the superior ability of the cotton fibres to absorb stress transferred from the matrix, compared to that of the pure geopolymer.

Fibre orientation in relation to the applied load was found to be an important factor affecting the mechanical properties of cotton fabric-reinforced geopolymer composites. The mechanical properties of those composites with horizontal fabric layers (i.e., load normal to fabric layers) were higher than those with vertical fabric layers (i.e., load parallel to fabric layers). This may be because of the ability of horizontally laid cotton fabric to directly absorb and uniformly distribute the load throughout the cross-section. In addition, this enhancement of strength in the horizontal direction is due to the absence of shear loading at the fabric-matrix interface, which reduces the possibility of fabric detachment or delamination from the matrix at high loads. When composites contain horizontal cotton fabric layers, the fabrics are stretched, and cracks develop through fabric layers in a graceful failure behaviour, leading to a high degree of ductility.

The contribution of the reinforcing cotton fabric to crack arresting and bridging was also observed. The crack propagates through the specimen from one fabric layer to the next. Such crack arresting, bridging mechanisms and crack deflection are believed to be responsible for the significant enhancement in fracture toughness. However, when fabric is oriented vertically to the applied load, fabric delamination and detachment from the matrix can occur, resulting in inefficient stress transfer between fabric and matrix. In addition, at vertical loading the cracks develop along the fabric layers mainly through the geopolymer matrix, leading to a more brittle behaviour.

The water absorption of CF-reinforced geopolymer composites at room temperature is increased with increasing fibre content. Reductions in flexural strength, flexural modulus, impact strength, hardness and fracture toughness are caused by exposing the composite to water for a prolonged period. The deterioration of mechanical properties is attributed to degradation of the bonding at fibre–matrix interfaces as an effect of water absorption. Water penetrates the cotton fibre bundles and decreases their coherence, causing them to break down into finer fibrils. SEM micrographs show extensive fibre pullout and no evidence or trace of the geopolymer paste adhering to fibre. The observation is indicative of poor fibre–matrix adhesion in wet composites. In contrast, prior to exposure to water, SEM micrographs show almost no fibre pullout, undamaged fibre bundles, and pieces of geopolymer attached to the surface of the cotton fibre. Such observations are indicative of strong fibre–matrix bonding in dry composites.

Exposure to high temperatures (200–1000 °C) also leads to reduced mechanical performance (compressive strength, flexural strength and fracture toughness) in cotton fabric-reinforced geopolymer composites. As the samples were heated to 200 and 600 °C, it was found that some of the cotton fibres survived (confirmed by SEM); however, once specimens were heated to 800 and 1000 °C, a severe loss in strength was observed. The explanation for this is that as the temperatures increase beyond 600 °C, the oxidation and consequent degradation of the cotton fibres causes the sample to become highly porous (as confirmed by SEM and optical images). Furthermore, networks of hairline cracks were observed on pure geopolymer samples heated to 400 and 600 °C. Beyond 600 °C, severe cracking occurred on the surfaces.

When cotton fibres were incorporated, no such cracks were found in the surface of geopolymer composites. This suggests that cotton fibres are effective in preventing cracking caused by high temperatures. This prevention is due to the formation of small cavities inside the matrix, created by the fibre degradation, which prevent crack propagation. In this case, the structure of the composite becomes more porous, and expanded water vapour escapes without substantial damage to the microstructure. In addition, all samples exhibited a change in colour after exposure to elevated temperatures as follows: grey at 200 °C, yellowish grey between 400 and 600 °C, pink at 800 °C, and reddish brown at 1000 °C. These changes in colour are a result of the oxidation and liberation of iron species present in the fly ash particles. Such colour changes will be a useful tool for estimating temperatures reached after exposure to fire.

It is concluded that cotton fibre in the short and in the textile woven form are effective reinforcements for geopolymers. The mechanical properties of woven cotton fabric-reinforced geopolymer composites are comparable with or superior to those of geopolymer composites with short cotton fibres. The geopolymer composite made of woven cotton layers exhibits greater load-bearing capacity than that with short cotton fibre composites. This is believed to be because of better wetting by the geopolymer paste, and therefore better fibre–matrix bonding. In addition, since the precise content of short cotton fibres in geopolymers composites is an important parameter affecting the performance of these composites, an advantage of impregnated woven fabrics with geopolymer paste is that the fibre content in geopolymer composites can be increased without the adverse outcomes observed in the short cotton fibre composites: thus, by increasing the number of cotton layers significant strength improvement is attained. Natural fibres in textile form, as in cotton fabric-reinforced geopolymer composites, can be regarded as a good alternative to synthetic fabrics. These natural fibre fabric composites, with their good mechanical, environmental and economic properties, are a suitable choice for load-bearing applications.

7.2 Summary on Flax Fibre Reinforced Geopolymer Composites and Nanocomposites

7.2.1 Flax Fabric-Reinforced Geopolymer Composites

Geopolymer composites reinforced with different loadings of flax fabric (0–4.1 wt%) were fabricated. The increase in flax fibres content in geopolymer composites was found to improve the mechanical properties of the composites. The flexural and compressive strength, flexural modulus, hardness and fracture toughness improved at a maximum fibre content of 4.1 wt% (i.e., ten layers of flax fabrics). The flexural strength of the composites increased from 4.5 MPa in pure geopolymer to about 23 MPa (about 410% increase), and the compressive strength was found to increase from 19.4 in pure geopolymer to 91 MPa after the incorporation of FF. A slight improvement was also noticed on the flexural modulus and hardness of geopolymer composites as compared to the neat geopolymer.

The fracture toughness of the composites was also improved after the addition of FF. For instance, while the fracture toughness of pure geopolymer was measured as 0.4 $MPa.m^{1/2}$, after the addition of 2.4, 3.0 and 4.1 wt% FF, the fracture toughness was increased to 1.13, 1.5 and 1.8 $MPa.m^{1/2}$, respectively.

Regarding thermal stability, the composite revealed a total weight loss of ~15% at 300 °C which indicates a degradation of FF inside the composite. A large amount of fibre degradation has occurred at this temperature. The porosity of geopolymer matrices allowed the oxygen to move in and cause degradation of the flax fibres at high temperatures. For that reason, it could be concluded that this composite system is only appropriate for service below 250 °C.

7.2.2 Nanoclay-Filled Geopolymer Composites

Nanoclay particles were added to reinforce the geopolymer at loadings of 1.0, 2.0, and 3.0% by weight. Results showed that the mechanical properties of all geopolymer nanocomposites have been enhanced due to addition of nanoclay. It is found that the addition of 2.0 wt% nanoclay decreases the porosity and increases the nanocomposite's resistance to water absorption significantly. The optimum 2.0 wt% nanoclay addition exhibited the highest flexural and compressive strengths, flexural modulus and hardness. The geopolymer nanocomposite containing 2.0 wt% nanoclay decreased the porosity by 7.1% and water absorption by 17% and increased the density by 11.4%, flexural strength by 24%, compressive strength by 23%, flexural modules by 25% and Rockwell hardness by 12.6% when compared to the control geopolymer sample.

The microstructural analysis results indicated that the denser structure of nanoclay-geopolymer was because of the filler effect of the nanoparticles, as well as the high amount of geopolymer gel produced in the nanomatrices. Using FTIR technique, the level of geopolymerization can be identified by measuring the ratios the height and size of geopolymer fingerprint peak the Si–O–Si stretching peaks of the nanocomposites to the pure matrix. It was illustrated that all nanocomposites had generally higher contents of geopolymer gel as compared to the control paste; however, the addition of 2.0 wt% of nanoclay had the highest level of geopolymerization among all samples. SEM observation confirmed the previous results and showed denser microstructure in the case of nanocomposites. The geopolymer nanocomposite also exhibited better thermal stability than its counterpart pure geopolymer.

This enhancement can be attributed to the filler effects and activation of geopolymer containing 2.0 wt% nanoclay in which this nanocomposite had more consolidated microstructure than others. However, the addition of more nanoclay (beyond 2.0 wt%) into geopolymer nanocomposite adversely affected the mechanical and thermal properties. This could be attributed to the poor dispersion and agglomerations of the high nanoclay contents which created more voids in the matrix.

7.2.3 Nanoclay-Filled Flax Fabric-Reinforced Geopolymer Composites

Geopolymer composites and nanocompsites were fabricated using flax fabrics and nanoclay particles. The total amount of flax fabric in each sample was about 4.1 wt% and different contents of nanoclay (1.0, 2.0 and 3.0 wt%) were used. All samples were fabricated and stored for 4 weeks before testing. The effect of nanoclay on physical and mechanical properties of flax fabric-reinforced geopolymer composites as well as flax fibre-matrix interface was investigated.

The results showed that the optimum content of nanoclay was 2.0 wt%. The flax fabric-reinforced nanocomposites containing 2.0 wt% nanoclay decreased the porosity and water absorption by 16.3 and 19.4%, respectively, and also increased the density by 4.4% and flexural strength by 32.4% when compared to the flax fabric-reinforced geopolymer composite. This finding was also confirmed by the flexural toughness indices of the nanocomposites. The composite reinforced with the optimum loading of 2.0 wt% nanoclay showed higher toughness indices than the composite without nanoclay by 58, 54 and 39% for I_5, I_{10} and $I_{failure}$, respectively.

SEM micrographs of flax fibre-reinforced nanocomposite containing 2.0 wt% nanoclay also supported its improvement as compared to other composites in terms of better fibre-matrix interface. All nanocomposites displayed denser matrices with lower number of unreacted fly-ash particles embedded in the matrices. However, in flax fabric geopolymer nanocomposite containing 2.0 wt% nanoclay, a smaller amount of unreacted fly-ash particles has been observed, and higher content of geopolymer gel has been clearly seen, which provided better adhesion between the flax fibres and the matrix. However, less such enhancement has been observed when nanoclay content exceeds the optimum amount. This could be attributed to the increase in porosity which decreased the bonding between fibres and geopolymer matrices, and thus the associated load-transfer capacity. This means that pores and voids in geopolymer material acted as defects that weaken the composites.

7.2.4 Nanosilica-Filled Geopolymer Composites

The effects of nanosilica particles on the physical, microstructural and mechanical properties of geopolymer matrices containing nanosilica at 0.5, 1.0, 2.0, and 3.0 wt% were presented. The nanoparticles were mixed with geopolymer in two methods, dry and wet-mixing method. All samples were stored before test for 4 weeks at ambient temperatures. The way of mixing was found to manage nanoparticles dispersion, and thus to change the mechanical and physical properties of the resulted nanocomposites in both cases. For example, Si:Al ratios increased with the addition of nanosilica in all pastes due to the amount of silica added to the system. The Si:Al ratio started from 2.29% at the control geopolymer and increased up to 3.57% in dry-mix nanocomposites and 4.10% in nanocomposites created using wet-mix technique. All wet-mix nanocomposites showed higher ratios of Si:Al as comparing to their counterpart dry-mix samples.

It has also been observed that loading nanosilica to geopolymers using both preparation methods improved the density and decreased the porosity and water absorption of geopolymers. In wet-mix geopolymer samples, the density enhanced by 7.6%, while the porosity and water absorption decreased by 16.2 and 21.5%, respectively. However, in the situation of dry-mixing method, the optimum addition was found as 1.0 wt% of nanosilica, which improved the density by 15%, and reduced the porosity and water absorption by 27 and 35%, respectively, when comparing to the control sample.

The physical structure was found to have great impacts on the mechanical behaviour of geopolymer nanocomposites as the flexural and compressive strengths results followed similar trends to the densities of all nanocomposites. The flexural strength of dry-mix nanocomposites containing 0.5, 1.0, 2.0 and 3.0 wt% nanosilica improved by 20.0, 28.8, 24.4 and 15.5%, respectively. Whereas the flexural strength of wet-mix nanocomposites improved by 8.8, 15.5, 22.2 and 13.3%, respectively, as compared to the pure geopolymer sample. The compressive strength results of geopolymer nanocomposites prepared through both dry and wet mix procedures indicate similar trends to that of the flexural strength. The compressive strength of geopolymer nanocomposites improved by 21 and 27% after the addition of 1.0 and 2.0 wt% nanosilica in dry and wet mix samples, respectively. The strength was found to be directly proportional to the nanocomposites densities and inversely proportional to the porosities; nanocomposites with higher densities demonstrated higher mechanical results.

Results also showed that the addition of nanosilica particles improved the microstructure and the strength of geopolymer nanocomposites regardless of the method of preparation. Nevertheless, nanocomposites prepared using the dry-mix method demonstrated superior physical and mechanical properties as compared to their counterpart's wet-mix samples. The optimum addition was found at 1.0 wt% nanosilica that prepared using dry-mix procedure, any further addition of nanosilica particles beyond the optimum weight of 1.0 wt% adversely affected the physical and mechanical properties. This might be attributed to the poor dispersion and agglomerations of the high nanoparticles contents that formed more voids in the matrices.

7.2.5 Nanosilica-Filled Flax Fabric-Reinforced Geopolymer Composites

Flax fabric-reinforced nanosilca-geopolymer composites were fabricated with ten layers of flax fabrics (4.1 wt%) and similar contents of nanosilca (0.5, 1.0, 2.0 and 3.0 wt%). The samples were tested at 4 weeks after preparation. The ability of flax fabric-reinforced geopolymer composites and nanocomposites to absorb energy is identified by the flexural properties of the composites.

The results generally indicated that the addition of nanosilica into flax fabric-reinforced nanocomposite improved the mechanical flexural properties of the composites. Furthermore, the flax fabric-reinforced nanocomposites containing 1.0 wt% nanosilica achieved the highest improvements. The flexural strength of the nanocomposites containing 1.0 wt% increased from 23.03 to 30.50 MPa, about 32.4% increase as compared to flax fabric-reinforced composite.

The ability of a composites to absorb energy is identified by figuring out the flexural toughness indices I_5, I_{10} and $I_{failure}$. Flax fabric nanocomposite including the optimum addition of nanosilica particles and prepared through dry-mix procedure exhibited similarly the optimum flexural toughness. Composites loaded with 1.0 wt% nanosilica showed the maximum toughness indices of I_5, I_{10} and $I_{failure}$ as 44, 125 and 146, respectively. This improvement could be attributed to the enhanced density and the high content of geopolymer gel in the nanocomposites that led to better adhesion bond between flax fibres and geopolymer nanocomposites.

SEM micrograph of the fracture surfaces of flax fibre reinforced geopolymer composite revealed poor adhesion bond between the fibres and geopolymer matrices. On the other hand, good fibre-matrix interface as well as denser microstructure was observed in flax fibre-reinforced nanocomposites with the addition of 1.0 wt% nanosilca. The improvement in mechanical properties along with fibres-matrices interfaces of flax fibre-reinforced nanocomposites containing 1.0 wt% nanosilica was attributed to the fact that the nanoparticles improved the microstructure of matrix through geopolymeric reaction and pore-filling effect. Therefore, good interfacial bond between the resulted matrices and the flax fibres was achieved.

7.2.6 Durability of Flax Fabric Reinforced Geopolymer Nanocomposites

The effect of different loadings of nanoclay and nanosilica particles on durability and mechanical properties of geopolymer and flax fabric reinforced geopolymer composites was studied. The medium to long term durability of the composites has been investigated in terms of flexural strength obtained at 32 weeks. The microstructure was investigated using XRD, FTIR and SEM. XRD revealed a new carbonation phase (Trona) on the surfaces of geopolymer samples at 32 weeks. The carbonation content was confirmed when aged samples was tested using FTIR technique. FTIR also showed that the reaction has carried on at a low rate consuming more OH groups and forming tougher material during the ageing period. The water content decreased to some equilibrium level during this period resulting in lower broad peak at 3400 cm^{-1}.

At 32 weeks, the flexural strength of nanocomposites improved slightly when compared to their values at 4 weeks. For example, the flexural strength of geopolymer nanocomposite containing 2.0 wt% nanoclay increased from 5.6 to 6.1 MPa (9.0% incease). Whereas in the case of nanosilica, the flexural strength of the optimum addition of 1.0 wt% nansilica improved the nanocompsite from 5.8 to 6.2 MPa (6.8% increase), when compared to their values at 4 weeks. This slight enhancement in the mechanical performance could be attributed to the slow reaction of free silica and alumina in the presence of Na^+ ions during the 32 weeks.

On the other hand, all flax fabric reinforced geopolymer composites and nanocomposites showed reduction in the mechanical performance after the aging period by varying degrees. After the aging period of 32 weeks, the reduction in the flexural strength of nanocomposites was found less than that of control sample composite. The flexural strength of flax fabric reinforced geopolymer composites decreased by 23%, while the strength of flax fabrics reinforced nanocomposites containing 2.0 wt% nanoclay and 1.0 wt% nanosilica decreased by only 13.7 and 10.3%, respectively. This may be attributed to the point that the nanoparticles consumed amounts of the alkaline solution, reducing the alkalinity of the matrices, and producing higher amount of geopolymer gel, thereby improved the density of the matrices and the fibre-matrix adhesion.

SEM micrographs showed that flax fibres in flax fibres reinforced geopolymer composites suffer more degradation than that in the nanocomposites. Based on these observations, the addition of nanoclay has great potential in improving the durability of flax fabric reinforced geopolymer nanocomposites during ageing.

Finally, it can be concluded that geopolymer matrices modified with both nanoclay and nanosilica particles displayed an enhanced microstructural, physical and mechanical properties, and reduced the degradation of flax fibres in geopolymer composites. The nanoparticles showed comparable results; however, composites reinforced with nanosilica showed slightly better results than those contained nanoclay particles.

7.3 Concluding Remarks and Future Directions

Despite the significant enhancement and improvement in mechanical properties of natural fibre-reinforced geopolymer composites, there is a need to continue investigating the potential of cotton, flax and other natural fibres and nanoparticles as reinforcement for geopolymers. The following research activities are recommended as future directions for geopolymer composites work:

- The use of coupling agents should be considered to improve the fibre/geopolymer interfacial bonding and to provide better mechanical and thermal properties of the composites.
- Fibres could be pressure-treated with a low-viscosity chemical solution that penetrates the amorphous regions of the fibre and bonds with available OH groups. This would likely to reduce the hydrophilicity of natural fibres and improve the water resistance of natural fibre-reinforced composites.
- Loading of nanoparticles beyond the optimum addition into geopolymer nanocomposite matrices has been observed to adversely affected the mechanical and physical properties of geopolymer nanocomposites and flax fabric-reinforced geopolymer nanocomposites. Further investigation is required to resolve the issue of nanoparticles agglomerations and to identify the best method of mixing to achieve good dispersion of the nanoparticles at high amounts in geopolymers.
- Since the main concern of the use of natural fibres in geopolymer composites is long term durability of the fibres and composites, additional studies relating to strategies are necessary to overcome the degradation of natural fibres.
- The medium to long term durability of flax fabric reinforced geopolymer composites and nanocomposites has been investigated under ambient condition. Other durability conditions such as long-term durability, dry-wet cycles and freeze-thaw cycle should be attempted.

- A combination of short fibres and woven cotton fabrics as a reinforcement of the geopolymer matrix should be studied. This technique would give better reinforcement than woven cotton fabric alone because the spaces between the fabric layers and geopolymer matrix will be filled with short cotton fibres.
- The degradation of cotton fibre under alkaline conditions should be investigated. This will help determine the effect of highly alkaline environments on the fibres, particularly with respect to any dissolving of the lignin and hemicellulose phases and the likely weakening of the fibre structure.